D0396414

MADE FOR EACH OTHER

MADE FOR EACH OTHER

THE BIOLOGY OF THE HUMAN-ANIMAL BOND

Meg Daley Olmert
Illustrations by Virginia Daley

A MERLOYD LAWRENCE BOOK
LIFELONG BOOKS • DA CAPO PRESS
A Member of the Perseus Books Group

 Printed in the United States of America. For information, address Da Capo Press, 11 Cambridge Center, Cambridge, MA 02142.

Designed by Pauline Brown
Set in 13.5 point Perpetua by The Perseus Books Group

Library of Congress Cataloging-in-Publication Data
Olmert, Meg Daley.
Made for each other : the biology of the human-animal bond / Meg Daley Olmert.
p. cm.
"A Merloyd Lawrence Book."
Includes bibliographical references and index.
ISBN 978-0-306-81736-6 (alk. paper)
1. Human-animal relationships. I. Title.
QL85.O46 2008
304.2—dc22
2008035408

Published as a Merloyd Lawrence Book by Da Capo Press
A Member of the Perseus Books Group
www.dacapopress.com

Da Capo Press books are available at special discounts for bulk purchases in the U.S. by corporations, institutions, and other organizations. For more information, please contact the Special Markets Department at the Perseus Books Group, 2300 Chestnut Street, Suite 200, Philadelphia, PA 19103, or call (800) 810-4145, ext. 5000, or e-mail special.markets@perseusbooks.com.

10 9 8 7 6 5 4 3 2 1

For Michael

Contents

	Introduction	ix
ONE	Just Watching	1
TWO	The Birth of the Bond	15
THREE	A Mind on Nature	33
FOUR	The Rules of Engagement	43
FIVE	Brave New World	51
SIX	Good Dog	67
SEVEN	On the Shoulders of Giants	81
EIGHT	The Meeting of the Minds	105
NINE	The Dog of the Hare	121
TEN	New Game	137
ELEVEN	Made for Each Other	151
TWELVE	The Survivors	165
THIRTEEN	The Kids in the Coal Mine	179
FOURTEEN	Oxytocin Deprivation	195
FIFTEEN	Just Realizing	219
	Notes	245
	Acknowledgments	277
	Index	279
	About the Author	284

Introduction

We live surrounded by miracles we call pets. This may seem like an overstatement when your dog is whining to go out on a cold, rainy night or you're writing that fat check to his vet, but it's not. The fact that wolves stopped stalking us and we took them into our caves proved to be a miraculous leap of faith that changed our world forever.

In fact, the animals we can't live without were once creatures impossible to live with. They attacked us or ran from us, and we hid from them. So what happened? How did it ever come to this?

Thousands and thousands of years ago, our ancestors dreamed of unions with animals that would make them stronger, braver, faster, and wiser. In their dreams they surrendered their humanity and took on the shape and power of wild beasts. It was these mergers with the animal form and spirit that humans believed to be their ultimate achievement. They knew it was only with the help of animals that they could navigate this life and cross over into the next.

But how do you meld with animals you cannot even touch? That was the dilemma our human ancestors faced for many thousands of years. The animal power they worshiped and coveted was designed to crush them—and did so all the time. The dream of animals as protectors and allies was just . . . a dream. Waking relationships with animals were more of a nightmare.

But the dream did come true. This book is about the science behind that miracle—the miraculous but decidedly natural story of how and why animals and humans can stop being enemies and even fall in love.

I have always been fascinated by animals, even though I did not know any intimately until I was an adult. I grew up in the heart of New York City, in a housing project that did not allow animals. My first exposure to pets was the fierce-looking guard dogs that patrolled the meaner streets bordering my neighborhood.

I was never frightened by those dogs and, much to my parents' horror, would run to them and bury my face in theirs. They never bit me and I somehow knew they wouldn't.

Adults would say to my parents, "She has a way with animals." I remember that clearly in the way children will often remember things overheard about them more clearly than said directly to them. I also remember it made me feel good, even if I had no idea what it meant to a city kid like me. That and a token would get me on the subway.

Eventually I got out of the city and found a job working in a veterinary clinic where my "way with animals" was remarked upon by people who actually knew what they were talking about. I began to wonder why I knew how to talk to and touch animals. I never saw it done, was never taught any approach or technique. The way I moved around animals just came to me as naturally as breathing.

I eventually moved away from my country life but never really left the animals behind. When I moved to Washington, D.C., I had the good fortune to get a job at National Geographic Television. There I was able to observe people who had a way with wild animals. I couldn't help wondering who first looked up at an elephant and thought it would be a good idea to climb on its back and poke it with a stick? What kind of person had the crazy idea to try to ride a horse instead of eating it? And why on earth did a horse ever allow such a predator onto its back? Clearly, those who first found they had a way with animals over time changed the nature of many animals, and the whole world for that matter. Once upon a time, having a way with animals must have been the most valued talent anyone could be blessed with. Did success with animals offer such a survival advantage that it actually entered the gene pool?

My musings on the nature of the human-animal bond took a dramatic turn in 1993, when I was hired by actor and conservationist Stephanie Powers to develop a series on the history of humans and animals. My research lead me to people past and present who had extraordinary ways with animals. I also came across scientists who were beginning to realize that animals can have an extraordinary way with people. These researchers were asking animals to treat people with troubled hearts and minds and finding that they would and could.

One conversation proved to be another turning point for me. It was a phone interview with Aaron Katcher, a psychiatrist with the Devereux Foundation in Glenmore, Pennsylvania. He was working with young boys who had severe forms of attention deficit hyperactivity disorder exacerbated by behavioral problems as well. He had paired these troubled boys with animals—rabbits, turtles, gerbils, doves—and found that the presence of these creatures made them calmer and more attentive.

Caring for animals helped them hear what others had to say. Their quieted minds could even detect the nonverbal signals essential to successful social communication. But what was it about the zoo pets that made these boys able to sit still and take notice? The novelty of the animals' behavior was one thing, Katcher explained, but it was the next thing he mentioned that caught my full attention. He told me about an earlier study of his showing how interactions with animals can lower a person's heart rate and blood pressure. I asked him what biological mechanisms animals were activating in humans to produce such dramatic physical and behavioral changes in us. But I was amazed to hear Aaron Katcher say that no one knew. No one had even really investigated it.

The next day I happened to read an article in the *New York Times* about a hormone called oxytocin, which induces labor contractions and milk ejection in new mothers. This article was about new research showing that oxytocin was also capable of creating monogamous social bonds and parental devotion in a species of rodent. Even more interesting to me, a study showed that the high levels of oxytocin in lactating women also made them feel calmer and more sociable. The more oxytocin flowing through them, the more they were inclined to ask for and listen to advice and also to be more sensitive to nonverbal communication. I could barely believe my eyes. This hormone seemed to be producing the same psychological transformations in new mothers that animals were inspiring in Aaron Katcher's kids. Did caring for pets release oxytocin in humans too? And what about the strong sense of connection and devotion that Katcher's kids—and all pet owners—feel toward their "babies"?

I called C. Sue Carter, a researcher mentioned in the article. She listened patiently as I explained Devereux's zoo program and the similarities in the effects of caring for animals and nursing babies. Then I asked if oxytocin might be causing the changes in both cases—and if so, did that mean that oxytocin might be the chemical behind the human-animal bond?

There was a moment of silence on the line before Dr. Carter slowly responded, "Yes." She said it was surprising for a nonscientist to make such connections, but she and some colleagues had been considering such possibilities recently. She said that, even though no one had yet investigated whether caring for animals could release oxytocin in humans, this surely must be the case.

Sue Carter invited me to her laboratory at the University of Maryland in College Park, where she was studying the biology of prairie voles, a creature with strong monogamous and parental bonds. That meeting lead to many others as Sue generously became my oxytocin mentor. She not only shared her research with me but also introduced me to her colleagues. She made sure I met the Swedish researcher Kerstin Uvnas-Moberg, who had discovered those amazing personality changes that oxytocin produces in nursing mothers.

I became further convinced that oxytocin was the biological force at work in Aaron Katcher's patients when Kerstin told me of studies in her lab which showed that giving oxytocin to rodents lowers heart rate, blood pressure, and levels of stress hormones. Gentle human touch can also release oxytocin—and its effects—in animals. She had produced these results herself by stroking rats rhythmically forty times a minute—the same timing and gestures we use to pet our companion animals!

Immediately I wanted to know if it worked the other way around—could contact with animals release oxytocin in humans? Again, I hit a scientific wall, because that was an experiment no one had conducted. With the help of Sue Carter, Kerstin Uvnas-Moberg, and another oxytocin expert, Dr. Cort Pedersen of the University of North Carolina, we designed a study to measure oxytocin levels in humans before, during, and after interactions with their pets. Much to my frustration, I could not find a research lab with the time, money, or interest to pursue it.

I never gave up the oxytocin trail, however. Over the years I collaborated on writing and research projects with both Sue and Kerstin, and I read every oxytocin research paper I could get my hands on. The more I understood about the role oxytocin plays in helping to form powerful social and parental bonds, the more I was convinced oxytocin could create the powerful emotion attraction that bonds humans and animals.

It was an exciting time to be oxytocin watching. Sue, Kerstin, and Cort were at the vanguard of a small cadre of scientists who were discovering that oxytocin was far more than the hormone of labor and lactation. These researchers were learning how oxytocin works within the main brain centers that control emotions and behavior. In fact, oxytocin is central to a mininervous system that can shut down the body's most powerful defensive system, fight/flight, and replace it with a chemical state that makes us more curious and gregarious.

Is this the chemical shift that inspired wolves to enter our caves and us to pet them? Once we touched them and began to care for them, did we unwittingly create a mutual flow of oxytocin between us that made us well up with affection for each other? Was oxytocin also the main biological ingredient of domestication? It all made perfect sense. I just needed proof that animals can trigger oxytocin's bonding chemistry in us.

In 2003 I got my answer. Two South African researchers, Johannes Odendaal and Roy Meintjes, measured the blood pressure and blood chemistry of eighteen humans and dogs before, during, and after they had friendly interactions. The results were better than I dared to dream. Not only did the blood pressure go down in both the humans and the dogs, their oxytocin levels rose significantly. In fact, the oxytocin levels in the humans and dogs almost *doubled*—making pets one of the most potent triggers of oxytocin production in humans.

With this last link in place, I was emboldened to write this book. I have done my best to weave together the scientific clues that I believe put oxytocin not just at the heart of the human-animal bond, but squarely behind the domestication of animals and even the domestication or civilization of humans.

These are big assertions, but every day more is learned about oxytocin that supports them. In the past two years, researchers have found that oxytocin nasal sprays are safe and effective when used to study its effect on human subjects. And they are finding that oxytocin does the same thing in us that it does in other animals. Oxytocin lowers heart rate and stress hormones. It makes people more trusting and more trustworthy. It can even relieve some of the antisocial tendencies of autistics. Sophisticated brain imaging techniques are allowing us to see the human brain on oxytocin. And what we see is that a whiff of oxytocin will quiet the brain's fear centers even when subjects are shown pictures of people making threatening faces.

Taken together, these most recent human studies and the thousands of animal studies that have examined oxytocin's physiological and behavioral effects begin to make a broader case. Oxytocin may have stoked the warmer social climate that emerged during the long, stressful Ice Age. The triumph of trust over paranoia enabled humans and animals to come together in domesticated partnerships and emboldened people to move beyond the social limitations of kinship and tribe and live harmoniously in a civilized world.

In the coming chapters you will meet modern-day people who still have the oxytocin stuff to make the wild beast mild. These are the rare humans who see animals as we once did. These are people who retain the kinds of heightened observational skills and preverbal eloquence that allow them to move among wild animals and to be seen by them as friends, even leaders.

You will also learn how the gentle, patient temperament of the first animal whisperers could have ignited biological forces that shut down the fight/flight reflex and encouraged animals to relax in the presence of a human. This same socializing biology flows through us all, and we've never been immune to its civilizing effects. For example, we know from the archeological record that as wolves morphed into dogs, humans were becoming kinder and gentler too. Humans were banding together to hunt, eat, and share the duties of raising our big-brained, helpless young. You will see how and why the oxytocin released during these kinds of nurturing behaviors could have made our ancestors begin to want to touch and care for animals as well.

When humans began to keep animals and animals submitted to our care, we inadvertently created a powerful chemical feedback system that changed our hearts and minds. This is the biological synergy that brought humans and animals together and is still flowing through hundreds of thousands of pets and their owners all over the world today.

Contrary to the romantic myth, these stunning emotional and therapeutic effects are not the product of our pet's "unconditional love" for us. The research described in this book pulls back that sentimental curtain to reveal the very real and even more wondrous science behind it—the physiological reality of why animals can love us, why we can love them, and why that love is so good for everyone it touches.

For me, this lesson came to life when a seven-pound cat named Princess saw something in me I didn't know existed. Princess was one of dozens of feral cats living short, brutal lives on my neighbor's farm. She came to me out of the blue while I was writing this book and offered her love and devotion. I have had many wonderful animal friends, but this was something different and everyone who knew us could see that. Soon they could also see that Princess was pregnant. It was when she had her kittens that the words came flying off these pages and hit me in my heart.

Princess and I hovered over her babies, stroking them, kissing them, and retrieving them from impossible nooks and crannies. It was during this busy time that I became privy to the secret language momma cats murmur to their babies. Long after the kittens had moved to new homes,

Princess and I would use these gentle grunts and mews to relay our pleasure and need for each other. Some special chamber of my heart was pried open, and I experienced a depth of devotion to that animal that knew no bounds. And she felt it too.

I hope all mothers love their children that much, but I suspect that what I felt is a love reserved for humans and animals. I can't say why Nature would have designed us to fit together so perfectly with another species—or why some people don't. But thanks to Princess, I do know—both intellectually and viscerally—that such a power exists. The human heart does beat differently for an animal, if we let it.

Since before we painted thousands of animals on rock walls across the globe, we sensed we were made for each other. This book will explain why that instinct was just as right then as it is now. It was always oxytocin drawing humans together and making us feel better in each other's company. We will follow this ancient chemical trail to its twenty-first-century conclusions. We will consider the future of our best friendships, those that were born in a vast natural landscape and raised on farms that steeped us in oxytocin for the last ten thousand years. As our landscape becomes more paved and polluted, it is satisfying to know that our uncontrollable desire to keep pets, as well as our nostalgic yearning for the family farm and our hope to preserve animals in the wild, makes good evolutionary sense. We give the animals under our care love and attention, and they return it in full, further casting us under the spell of this powerful antidote to modern-day stress.

In the end, our big brain may just be big enough to grasp this deeper understanding of the biology of the human-animal bond. We may even be smart enough to accept that exposing its humble molecular roots does not diminish or reduce its life-changing powers, but opens them up to a new level of exploration and admiration. This is science that supports a truth the heart has always known.

ONE

Just Watching

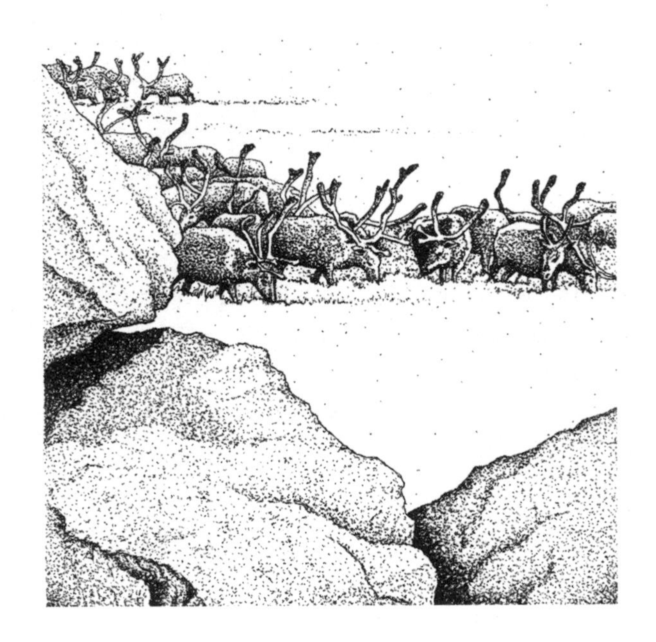

In the beginning we just watched. Incredibly, the profound bond we share today with animals grew out of a glimmer in the eye of the earliest humans. These were members of the bipedal species of primates who walked away from the great African forests that slowly withered during the Ice Age. Stripped of their leafy refuge, these upright apes fanned out into the grasslands that widened before them and dared to match their wits against the legions of mighty animals that ruled the Pleistocene.

Some took it harder than others. Some scanned the bright, open landscape and saw opportunity in the danger. Some even figured out how to grab it. They were the hominids who learned to see the world as no other primate had before—the ones who became social carnivores.

Communing with animals was far from the minds or hearts of these pioneering primates. Their only desire was to avoid some animals while trying to figure out how to eat others. They faced this unlikely challenge with highly developed visual, motor, and vocal/auditory skills. Their brain had been forged in a dark and tangled world as well as a complex social one. But could the largest brain on earth cope with the harsh reality of the savanna?

Foraging had never looked like this before. Dinner was not going to be found lurking under a leaf or high on a limb. Stomachs that ached for fruit would now have to be quieted with roots, bugs, honey, eggs, worms, and the occasional rodent or reptile. These morsels must have seemed paltry compared to the big game that predators were gorging on in the distance. But those feasts belonged to natural-born killers with the fangs and speed to claim them. Our ancestors lacked this physical might, but they possessed sharp eyes and a social brain that could see the world through the eyes of other creatures. It was a challenge that would expand their minds as well as their diets. Philosopher Paul Shepard explains it this way:

> As we became open country hunters and hunted, we entered an on-going system of brain-making, using our advanced primate vocal and visual apparatus instead of an olfactory system. This process centered visual imagery on the simultaneous and *tireless scrutiny of other animals* and the emergence of self-consciousness. [my italics]

Shepard proposes that for hundreds of thousands of years, animals made a big impression on the growing minds of our savannah-dwelling ancestors. It was animals that terrified, awed, delighted, clothed, fed, sheltered, and mystified them. Shepard thinks we sculpted our great cauliflowered brains while staring at them.

Here Shepard is taking a page from paleoanthropologist Harry Jerison, who in *Brain Size and the Evolution of Mind* shows that the pressure to devise avoidance and attack strategies fostered mutual brain growth in carnivores and in the ungulates they ate. In their quest to become carnivorous, hominids faced the double jeopardy of being both predator and prey, and they did it without the physical talents to outrun or overcome other animals. But they did share one key trait with the animals surrounding them—they too were social. Those who paid the closest attention saw how other animals increased their strength and safety through cooperation. This was one survival strategy that early humans could employ and did. They formed new alliances that helped them run the deadly predator-scavenger gauntlet and they learned to share the stolen rewards with their co-conspirators.

One creature's trash has always been another's treasure, and these prehumans found their prize waiting in the bones and skulls of abandoned carcasses. These remains happened to be filled with fatty acids—just the stuff that grows bigger, better brains. And so with their clever hands they fashioned tools, split open the skeletons, sucked them dry, and truly "entered an on-going system of brain making."

It is very hard for us to fathom a world in which we were not top predator. But to get a sense of what filled our brains before language, we need only look back a little over one hundred years ago to places where such Paleolithic realities still existed.

In 1892 American frontiersman and conservationist C. J. Buffalo Jones witnessed one of the last demonstrations of the natural order that shaped the logic of our ancestors. On a hill in Canada's Northwest Territory, he watched herds of caribou congregate until all the eye could see for ten miles around was a giant mass of animals whose antlers became a mighty forest. For several days this living landscape flowed past him day and night. Jones halved his estimate on the herd's size to a conservative 25 million, but conceded that "it is possible that there are several such armies."

That's a lot of animals, too many for our eyes to count or believe, but it was the reality once upon a time—a time when there were only a million humans hunting and scratching out a living in the world. Experts estimate that a classic "herd" of Paleolithic humans numbered a couple of dozen, and only seven of those were adults. On special occasions—such as periodic ritual celebrations or great hunts—the human herd might swell to as many as five hundred. So it was a wildly outnumbered human brain that was asked to absorb the implications of animals as superorganisms as well as discern the minutiae of their individual behavior and anatomy.

Our ceaseless need to assess the strengths and weaknesses of animals, to cope with their overwhelming presence or threatening absence, was a matter of life and death. But analyzing animals proved to be more than a survival strategy; it became school, television, even church. It is no wonder that a history of staring at animals has left us with a brain that still can't help but seek them out and try to understand them.

"Tireless scrutiny" showed our ancient ancestors how to scavenge the dregs of big game and eventually to get the lion's share. But what if these earliest humans weren't just watching animals for all those hundreds of thousands of years? And is there even such a thing as "just" watching?

Fifteen years ago, researchers at the University of Parma in Italy began to scrutinize how the primate brain plans and executes movement. To do this, they implanted electrical sensors in a monkey's motor cortex to see which areas were activated when the animal picked up or put down an object. One day the brain monitor went off, indicating that the monkey was moving his arm—only he wasn't. The monkey was sitting perfectly still *just watching* a lab assistant eat an ice cream cone. The buzzing electrodes had exposed one of the brain's oldest secrets: that just watching another person or animal perform an action causes our brain cells to fire as if we are actually making the movements ourselves. Dr. Rizzolatti's team had stumbled across an amazing link between vision and action taking place in specialized brain cells they called "mirror neurons."

In humans, mirror neurons also resonate with the thoughts, intentions, and feelings of others. Thanks to the reflective power of mirror neurons, our brains, at a glance, can understand if a person picking up a glass is going to drink from it or throw it across the room in anger. They help us feel that anger as well. Mirror neurons are considered the most important missing link in the evolution of mind because, as Rizzolatti showed, "they allow us to grasp the minds of others . . . by feeling, not thinking."

And because conscious thought is not required in order to experience this imitative effect, it seems logical that mirror neurons helped early preconscious humans internalize the life-and-death dramas they witnessed constantly. The synchronized thrust and parry of predator and prey that our hunting ancestors mastered certainly implies a deep reflective undercurrent at work. The discovery of these see/do/feel brain cells means that, even from a safe distance, eye contact was linking us in profound ways to other animals, because, Rizzolatti explains, "mirror neurons translate visual information into knowledge."

There's another curious thing that happens to the subconscious brain while we are "just watching." In the late 1960s psychologist Robert Zajonc showed that when humans or animals are repeatedly exposed to stimuli—living or nonliving, a member of the same species or not—they develop positive feelings for the stimuli. Zajonc called this most basic kind of bonding "the "mere exposure effect." Given the endless exposure our ancestors had to animal "stimuli," it is entirely possible they could have fallen prey to the mere exposure effect and been infused with a dawning appreciation of animals.

When you factor in the connective power of mirror neurons and the mere exposure effect, "just watching" can hardly be considered a passive, one-sided form of contact. Our attentive, watchful ancestors were absorbing the actions of animals, intuiting their next move, sensing their emotions and pain. These primitive systems of attention could have sparked in the growing human heart and mind a rudimentary sense of connection to animals long before we had the consciousness or language to discuss it or celebrate it.

Harry Jerison pokes another hole in the "just watching" notion. He believes that some of the things we stared at may have also elicited an automatic vocal-audio response. He proposes that the expressive brain wiring our primate ancestors relied on to signal danger or to mark their territory found new purpose in helping hominids navigate their vast and dangerous landscape.

Jerison points out that the ecological niche clever apes walked into belonged to animals who could scent mark their way around, smell blood, or sniff the slightest whiff of danger. Early humans had no such olfactory talents and had to glean all this information differently. Jerison thinks they absorbed their world by burning it into their retinas while taking vocal notes as a kind of aide-mémoire.

> It is an odd picture, but I think it works. Instead of urinating and sniffing . . . we can imagine our ancestors as marking with sounds and sensing the sounds—talking to itself, as it were, but in primitive tongues.

By vocalizing their visual world, Jerison thinks, they talked themselves into it in a new way. Just as with the mirror neuron response, this may have brought the outside world inside the human head and made it easier to remember, and eventually to think and talk about. Jerison proposes that these audio images eventually merged into an internal model of the world, planting it and all its endless creatures into our newly emerging sense of consciousness and language. So much for "just watching." Even from afar, our watchful ancestors were infused with a physical and psychological experience of a world defined by animals.

The urgent business of sorting and integrating the sensory experiences of the Ice Age proved to be just the sort of brainteaser that fosters neural growth. The artifacts of ancient hominid skulls leave hard evidence of a dramatically expanding neocortex. They tell us that between 2 million and 700,000 years ago, brain size doubled. This brain gain would make the various species of Homo erectus more cooperative and adventurous. They entered into alliances for hunting, gathering, and child care. They used fire and may have gone from muttering to themselves to muttering to each other in a kind of protolanguage. Some even migrated out of Africa for the first time. Then between 500,000 and 100,000 years ago, another spurt of brain growth gave us Homo sapiens, our direct ancestors. The skull had maxed out, leaving them (and us) with a well-fed neocortex three times the size of a chimp's.

The fossil record hints of the brain storm brewing in these sapient skulls. Complex tools appeared, things were traded widely, hunting and foraging skills gained sophistication. This is the brain that found its way out of Africa again and across Europe and Asia. This is also the brain that erupted forty thousand years ago in an artistic outpouring that left cave and rock faces on every continent covered with 45 million carvings and drawings.

And what was it these human hunters were finally compelled to draw? The vast majority of their images are animals. Animals, it seems, were the main muse to our artistic, carnivorous ancestors—with human body parts coming in at a distant second place as worthy subject matter.

Animals clearly left their imprint on our subconscious as well as newly emerging conscious brain. We may never figure out when or how we became conscious, talking beings, but we do know that when thoughtful conversations began, they would have been largely about the same old thing that preyed on everybody's mind: animals.

The record of our age-old obsession with animals can still be found in the developing brain of every newborn child. Human babies come into the world with a brain that is dramatically underdeveloped. Even at age two, a child's brain is still only 75 percent wired, leaving it more Neanderthal than Homo sapiens. Much of the neural work that remains to be done involves the establishment of vast communication networks between the left and right hemispheres of the brain—an evolutionary process that rolls out over the next twenty or so years.

Within minutes of birth, however, newborns can focus their primal attention systems on the business of finding the food, warmth, and attention that will keep them alive. Using the most primitive brain systems, babies learn their mother's voice, odor, and face. Their primate visual system, complete with mirror neurons, immediately kicks in, prompting them to mimic the facial expression and hand gestures of their caregivers—endearing them during this time of great vulnerability.

Yet in the midst of this eternal and essential search, there is still something strong enough to pull a baby's head away from the hand that feeds it—animals. The ancient brain centers that help babies recognize and understand face and body movements don't seem to discriminate between human and animal. And so young children delight at the sight, sound, and touch of animals, stuffed or otherwise. They follow them with their eyes, reach for them with their hands, and readily imitate their sounds with ease. This early keen interest in other life forms is a recapitulation—a vestige—of our ancestors' essential struggle as both prey and predator.

If visual fixation on animals helped groom our primitive brain for language, it's only natural that the animals we stared at might play a critical role in jump-starting the human infant's speaking career. Animal names and body parts—those same things that fascinated our predecessors hundreds of thousands of years ago—are among the first nouns babies delight in saying. Stephen Kellert, who studies how animals affect human health and behavior, has found that "animals constitute more than 90 percent of the characters employed in language acquisition and counting in children's preschool books." More than 90 percent! It's hard to imagine accomplishing the monumental feat of learning language without the inspiration of animals.

Language is probably the best trick of our human brain. Once we learned how to manipulate words with grammar, we were able to explore a vast new landscape of thought that stretched beyond the here and now. We entered into the internal dialogues that expand our consciousness. And we found the way to share those thoughts and remember them better. Some consider the emergence of language to be the brain's "big bang."

Despite the protestations of many linguists and philosophers, language did not so much separate us from animals as reorganize our relationship with them. Millions of years of visual contact with animals had forged the powerful nonconscious bond we felt with them. That visually inspired connection only deepened when our conscious brain gave birth to our sense of self-awareness and language. But it was to animals that we instinctively turned as we cast about for ways to feel more comfortable with our newfound powers.

The great speaking brain, after carefully discriminating between ear and nose, dog and cat, begins to play with their similarities: pig-nosed, dog-eared, lazy cow, busy bee. In fact, metaphor may have been the first job our big talking brains cooked up for animals. Metaphor, according to veterinarian and anthropologist Elizabeth Lawrence, is a symptom of the human need to affiliate with the rest of nature. "Through such symbolizing," she says, "there is a kind of merging—animals take on human qualities and humans take on animal qualities."

As we stared at animals, we not only figured out how to survive but began to recognize ourselves through exploring the similarities and differences between them and us. We looked into the animal world and found a curved mirror filled with our own reflection. In that mirror we gained our self-awareness and, from that, our culture. A way with animals became a way with words. So in a very real sense, the "big bang" of language did not separate us from animals but wove them ever deeper into our conscious and subconscious.

Animals were just this essential to the evolution of language, and consequently to human consciousness. Even in modern life where people's everyday experience of animals has dwindled, their presence in the deep recesses of our brains remains large and potent. Safely tucked away in our psyche, the ancient bond that was forged at a distance springs back to life unchanged and accessible to children young and old.

Edward O. Wilson, America's foremost evolutionary biologist, is not surprised to find animals so deeply embedded in human language and consciousness. Like Jerison, he recognizes that our big brains are not the product of the modern, technologically mediated world, but are shaped and informed by a more ancient and organic one. He believes millions of years of total submersion in nature left its mark not only on brain function but on our DNA. Consequently he believes we are delivered into this mechanistic world with an "innate tendency to focus upon life and other lifelike forms, and in some instances to affiliate with them emotionally." This ancient attraction to other living things he calls *biophilia*.

Wilson put his reputation as a serious scientist on the line when he first proposed his biophilia hypothesis in 1984. He readily admitted that "by ordinary standards of natural science the evidence remains thin, and most of the underlying theory of its genetic origin is highly speculative." Even so, he argued, "the logic leading to the idea is sound, and the subject too important to neglect."

Despite Wilson's prominence, his notion—that we are innately, genetically predisposed to be fascinated by the living world around us and to feel a strong attachment to it—met with mixed reviews. Some argued that it is our culture, not our genes, that determines what we find interesting or pleasing. One critic summed up the skepticism by saying that "genuinely universal needs are hard to find."

Wilson's hypothesis lacked a genetic smoking gun, and yet how could this serious scientist have reached any other conclusion when, at the age of nine, young Wilson was overcome by the urge to stare at ants? Beneath the bark of a rotting tree he happened upon a seething colony of citronella ants. So complete and instantaneous was his attraction, that he later concluded it could only have been innate. "Most children have a bug period," Wilson says, "and I never grew out of mine."

In his autobiography, *Naturalist*, Wilson talks about the moment in childhood when biophilia begins to take hold. In nature, Wilson says, children are exposed to classical, archetypical experiences that have proved to be valuable to our well-being throughout our long prehistory. The sensations felt during these encounters are not consciously articulated but leave the child with something wordless, something Wilson calls "guiding emotions."

The emotions that guided Edward Wilson to watch ants also led him into a kind of meditative state that he calls the "hunter's trance." He says he first felt the hunter's trance come over him as his eyes scanned the forest floor looking for his tiny subjects. He felt himself entering into an intensified concentration in which heart, breath, and mind are quieted. During this biophilia mind/body moment, Wilson says, the living world reveals its deepest and smallest secrets.

Nature's ability to turn our heads and hearts in its direction supplies an emotional and physiological component that sharpens our observational skills and makes us feel even closer to the living world around us. These "guiding emotions" are the time-tripping visceral expressions of biophilia that have ushered us from ape to human and into world dominance. In fact, the keen interest in the living world they promote has proven to be one of the most valuable pastimes of our big brains.

Testimonies supporting the existence of a preprogrammed attraction between living things are endless. Tens of millions of pet owners, for instance, will swear they feel as bound to their pets as they do to their families—if not more. But all these poignant stories of deep devotion and interspecies understanding do not a scientific case make. They remain warm and fuzzy anecdotes, mere ripples on the surface, hinting at greater forces at work beneath.

Still, our affinity for other life-forms is *not* just in our imagination or the way we speak—there *is* proof that we're born with it. In the pages that follow, I hope to show that our curiosity about other living things—as well as the urge to pet a dog or ride a horse—is biological, is genetic, and can stand up to scientific scrutiny.

Coincidentally, at just about the time Edward Wilson was formulating his biophilia hypothesis, researchers were discovering the affiliative powers of a hormone called oxytocin. It seems we're all born with this hormone and the urge to connect with others that it inspires. Oxytocin, as we will see, is what makes us not just mammals but *social* mammals. It is also a key ingredient of the chemistry that creates the human-animal bond, and it happens to be released in the brain while we are "just watching."

TWO

The Birth of the Bond

The quest for meat urged our predecessors one step closer to other animals. Hunger necessitated a new focus on creatures who had mastered the savannah. Tireless scrutiny of these social carnivores—the big cats and wolves, the hyenas, and the wild dogs—got us out of our primate-centric heads and into the minds of other animals. We came to recognize strengths to be mimicked and weaknesses to be exploited. But something else was happening deep in our brains as we watched animals. Emotions were stirring that would leave us feeling a profound sense of connection to them.

As soon as we could, we declared our fixation with animals by drawing them over and over again for tens of thousands of years. The human figure rarely appears in the midst of all these Paleolithic animals, but when it does, it's often represented with the body of a human and the head of an animal. These hybrid figures seem to say we had gone from a species that avoided animals to one that wanted to merge with them. A sense of attachment to animals was growing that would shape the future for us both. This attachment has its roots in the biology of reproduction. The investigation into this biology has only begun, but already one chemical stands out as its star component—an ancient molecule called oxytocin.

Oxytocin was first identified in 1902 as a hormone found in the pituitary gland that acts on the muscles of the uterus to produce labor contractions. This earned the molecule its name, which in Greek means "quick birth." It took scientists another fifty years, however, to realize it also produces the spasms that release breast milk. Oxytocin's critical assistance in labor and lactation sealed its identity as a "female reproductive hormone" until the early 1980s, when scientists began to realize that there was more to the oxytocin story.

All mammals, male and female, make oxytocin, and in other species it also facilitates delivery and breast-feeding. In the 1950s scientists studying rat brains discovered that oxytocin is not actually made in the pituitary gland but is released into the body via this portal at the base of the brain. It is synthesized just above the pituitary in a cluster of nerve cells called the hypothalamus—the brain center that keeps our organs and bodily functions in good working order. Then, in the 1970s, researchers discovered that oxytocin-producing cells also branch out from the hypothalamus into every key brain region known to be involved in behavior and emotion. The discovery of this extensive oxytocin brain network suggested to some scientists that this female hormone might be doing a great deal more than delivering babies and breast milk.

In 1979, two psychiatrists at the University of North Carolina, Cort Pedersen and Arthur Prange Jr., were the first to investigate whether or not oxytocin's interbrain connections also produce the powerful sense of attraction and nurturance that overcomes females after they give birth. They treated virgin rats with high levels of estrogen to simulate the state of pregnancy. Then, just before putting rat pups in their cages, they injected the brains of these faux-pregnant females with oxytocin. The virgin females quickly showed full maternal concern and care toward the pups, licking them, retrieving strays, even trying to nurse them.

To fully appreciate what oxytocin had done to these sexually naive animals, you need to know that virgin rats normally avoid offspring, possibly because they are frightened or repulsed by the sight and smell of naked, wriggling rat pups. Pedersen later led a team that confirmed oxytocin's power to induce maternal behavior by injecting natural rat mothers with a drug that blocks oxytocin's brain actions. Without the help of oxytocin, the rat mothers showed no interest in caring for their pups. These initial investigations revealed that, working within the brain, oxytocin can turn a female rat's fear and repugnance of pups into a magnetic attraction for them. Once female rats have been chemically convinced that pups are creatures to be nurtured rather than avoided or attacked, they will even care for strays that may appear in their nest.

Mother sheep, however, are far more discriminating about who gets their milk and attention. Because a herd of sheep may have many hungry lambs in it, mother ewes must develop a more specific recognition of their own baby. If a ewe fails to recognize a lamb as her own, she will head-butt it away, denying it the milk and protection it needs to survive. During labor, the ewe's brain must acquire the ability to distinguish the cry, smell, and face of her own lamb from all others. The birth process, in fact, is essential to the formation of lamb recognition and the onset of maternal behavior. Only female sheep that have just given birth will show any interest in nurturing lambs.

In the late 1980s, Keith Kendrick and E. Barry Keverne were part of an international team of neuroscientists trying to understand what happens during sheep parturition that helps a ewe recognize and care for her lamb alone. It was well known that oxytocin's presence in the uterus and cervix stimulates labor contractions, but Keverne and Kendrick also found that during labor, oxytocin levels rise dramatically in the ewe's cerebrospinal fluid (CSF). They also discovered they could trigger this spike of brain oxytocin by manually manipulating the vagino-cervical region. They wondered if this labor-induced appearance of oxytocin in a ewe's brain fluid might be what stimulates the lamb-learning factor so critical to a sheep's maternal behavior. As Pedersen and Prange had done in rats, they injected oxytocin into estrogen-primed, nonpregnant sheep, and these normally disinterested females suddenly displayed the full range of maternal recognition and acceptance of the lambs presented to them.

This realization—that within the brain, oxytocin is capable of producing the calming and cognitive prerequisites that allow a mother to identify with and attach to her baby—opened up a whole new appreciation of oxytocin and the biology of bonding. It inspired some intrepid researchers to begin exploring the possibility that oxytocin and other brain chemicals might be equally important in establishing the many other kinds of social attachments we mammals make.

Oxytocin's socializing influence was particularly intriguing to those researchers studying the rare social bonding ability of the prairie vole. This small brown rodent, native to the grasslands of the Midwest, is exceptional because it mates for life. Voles are highly social animals who live in extended family burrows organized around a main male-female mating pair. The pair share parental duties, and approximately 75 percent of their offspring remain in the family burrow to help care for new sibling litters.

Mating for life is for the birds, so to speak. Ninety percent of birds form these enduring partnerships, but even bird bonds have their limits. DNA tests have recently revealed that the staunch loyalty birds display toward their mates doesn't necessarily include sexual fidelity. Still, they remain dedicated to each other and "their" young, sharing the chores of feeding and protecting them. Scientists have a term for this kind of blind-eyed devotion—"social monogamy."

The lifelong commitment of the prairie vole is also more social than sexual, but even this is rare for mammals. While maternal bonding is essential to the survival of all mammals, other social commitments vary in necessity and degree, with less than 5 percent of all mammals forming lasting pair-bonds and sharing parental duties. Even closely related vole species, such as the montane vole, have no such social inclinations. They live solitary lives and come together only to mate. Females raise their young alone, abandoning the nest as soon as possible.

C. Sue Carter was one of the first researchers to investigate the biology behind the prairie vole's monogamous bonds. In her lab at the University of Maryland in College Park, she and her team discovered that males and females placed in the same enclosure would approach each other for an initial exploratory sniff before settling down side by side for hours. After this simple form of extended contact, the pair elected to stick close to each other even when new animals were added to their cage.

Sitting together for hours also brought the female into a state of sexual receptivity that resulted in enthusiastic mating bouts over the next forty-eight hours. Afterward, the pair continued to choose each other's company over the social advances of other voles. Once prairie voles recognize each other as mates, they can hold that image forever. Sue Carter's field studies show that even when one member of a pair dies, there is less than a 20 percent chance the surviving mate will bond again.

Carter was aware of rat and sheep studies linking oxytocin to maternal behavior and selective bonding. She wondered if oxytocin might also be involved in her subjects' preferential and enduring bonds. She and her team infused the brains of female prairie voles with oxytocin before putting them in cages with males. Even without social stimulation, the females developed a distinct preference for their new cage mate. The scientists then gave these smitten females a shot of an oxytocin-blocking drug, and the bonding spell was broken. Carter and her postdoctoral fellow, Karen Bales, also showed that oxytocin injections encouraged young, sexually inexperienced female voles to lick and care for pups placed in their cage.

Carter's team found that oxytocin injections had similar effects on male prairie voles. After treatment they became dedicated to the female in their cage and showed paternal care toward pups presented to them. But there was one important difference. Normally, once attached to a female, a friendly, docile male will ferociously defend his mate from all suitors. The oxytocin injections did not bring out this aggressive behavior in the newly bonded male. Carter was not completely surprised by this finding, since there was nothing in the oxytocin literature linking it to aggressive behavior. She suspected that the male prairie voles' monogamous traits might be inspired by a closely related brain hormone called vasopressin.

Vasopressin and oxytocin have genes that are close neighbors on the same chromosome. They are made in the same nuclei of the brain's hypothalamus. Both are produced from nerve endings that reach throughout the mid- and lower brain, and both are released into the bloodstream through the pituitary portal. Their physical resemblance is so strong (differing only by two amino acids) that they can attach to each other's receptors, a capacity that allows them to trigger each other's effects. Most the time, however, they produce opposite behavioral and physiological reactions that bring each other into a healthy balance.

Early vasopressin studies showed that it's released during mating and can cause aggressive and territorial behavior. Carter found that vasopressin treatments cause male prairie voles to selectively attach to a female and fight to defend her.

This new understanding of the role played by oxytocin and vasopressin in the creation of pair-bonds and parental behavior in prairie voles begged the next question. If all rodent species make the same amount of these brain chemicals, why aren't they all monogamous? Dianne Witt, from Sue Carter's lab, and Thomas Insel's team at the National Institute of Mental Health in Poolesville, Maryland, were the first to discover that monogamous and nonmonogamous rodents have big differences in where their brains produce receptors for oxytocin and vasopressin.

Receptors are the proteins on brain cells that receive chemical signals and respond to them. Female prairie voles, it turns out, have far greater densities of oxytocin receptors in the neural circuitry that supports something called "social recognition"—the ability to distinguish baby, kin, and friends and to remember them. Social recognition is essential to the formation of all social bonds. Male prairie voles make more of a particular vasopressin receptor—V1aR—in their social brain centers.

The brains of nonmonogamous female rodents only produce a monogamous-looking oxytocin receptor pattern right before birth. This flush of oxytocin receptors lasts a few weeks, allowing them to offer the maternal care necessary to keep their pups alive. When it fades, these females are gone and the weaned pups are on their own.

By the late 1990s the study of the behavioral effects of oxytocin and vasopressin had become a hot topic. Thomas Insel moved to Emory University in Atlanta to direct its new Center for Behavioral Neuroscience. There, he and Lawrence Young began experimenting with genetically engineered animals to look deeper into the mechanisms behind the bonding effects of oxytocin and vasopressin. They found that when they transferred the gene that makes the V1aR receptor from a monogamous male prairie vole into a nonmonogamous house mouse, the mouse's brain produced these vasopressin receptors in a more prairie vole–like way. The sudden appearance of V1aR receptors in cells that promote social behavior didn't make the mouse fully monogamous but much more social.

The team next took a gregarious species of mouse and made it less so by removing the gene to make oxytocin. These oxytocin "knock-out" mice could no longer make friends because they'd lost their ability for social recognition. The oxytocin deprivation had no effect on their capacity to learn other things such as the twists and turns of a maze, but their social memory had gone blank. When placed in an enclosure with another mouse, all the sniffing in the world couldn't help them learn the sight or smell of the other animal. When separated and reunited, the knock-out mouse sniffed away, totally unaware that this was the mouse it just met. Oxytocin injections restored their ability to recognize other mice, remember them, and be comfortable socializing with them.

A few years later, Young and his colleagues reported that vasopressin receptor knock-out mice also suffered social amnesia. With no receptors to respond to their vasopressin, gregarious male mice showed the same social blindness as oxytocin-deficient mice. These males also had no trouble with nonsocial memory tasks.

Rodents and sheep have given us tremendous insight into the complex chemistry that can turn strangers—even enemies—into lifelong mates, devoted parents, and just plain friends. But is this the same chemistry that inspires humans to feel strongly bonded to friends and family? Could this biology also be responsible for the human-animal bond? These questions are not so easily answered. Ethical and practical concerns prevent the kinds of brain-invasive research that has told us so much about the biology of bonding in other mammalian species. But there are recent DNA studies that suggest how these chemicals could be luring us into bonds with other creatures.

Larry Young and Elizabeth Hammock spotted a variation in a section of repetitive genetic material that appears just above the gene that regulates the production and distribution of the V1a vasopressin receptor. Monogamous rodents, like the prairie vole, make a long version of this "allele." In rats and house mice this allele is shorter, and montane voles have almost none. Young and Hammock have shown that, even within a species, a difference in the length of this allele could be correlated with just how social the individual is. Male prairie voles, for instance, with the longest alleles have the greatest number of V1a receptors in their social brain centers and show more loyalty to their mates and offspring than males with a shorter length of this mysterious DNA.

Young and Hammock have also found similar genetic differences in primates—including us. We humans and the highly affiliative bonobo apes have similar lengths of this allele just above our vasopressin receptor regulatory gene, while the chimpanzee has a shorter version. Young and Hammock surmise that this highly unstable genetic material may provide a kind of adaptive "tuning knob" that can quickly adjust social behavior to meet shifting environmental demands. The oxytocin gene also has a similar variation in the genetic sequences above the gene that regulates its receptor, but the effects of this differentiation on social behavior have yet to be investigated.

Still, the role of oxytocin in human social bonding is not a total mystery. Fortunately, there are times when what goes on deep in our brains makes itself abundantly clear. This occurs most dramatically in the physical and mental transformations that take place during and after childbirth that forge our first and most profound social bond. To understand the biology of the human-animal bond, we must begin by looking at the first social bond we humans make—the one between a mother and her baby.

Pregnant women, like all mammals, experience a sharp rise in their oxytocin levels just as they go into labor. This flooding of oxytocin—greatly stimulated by a concurrent rise in estrogen—produces the contractions that push the baby through the birth canal and out into the world. Once the baby is delivered, the sight, sound, smell, and even thought of her baby will cause the mother's hypothalamus to produce powerful electrical pulses that send oxytocin flowing into her breasts. The waves of oxytocin stimulate the nerve endings in the nipple to produce minicontractions that release breast milk into her baby's hungry mouth. The baby's natural suckle response also excites other breast nerves that send signals back to the hypothalamus keeping the oxytocin and breast milk flowing.

This elegant symbiotic design keeps babies fat and happy and nursing mothers awash in oxytocin. It is from these oxytocin-rich mothers that we have learned so much about the chemistry that makes us the kind of mammal that can form deep and long-lasting emotional attachments to our babies, our friends, our mates, our pets, even our planet. In humans, the story of bonding begins with the story of oxytocin.

Not all the oxytocin made in a mother's brain goes to her breasts. Her baby's suckling also stimulates a wave of oxytocin that spreads throughout her central nervous system to produce various neural reactions that change both her body and her mind. For instance, some oxytocin nerves reach into the brain stem and manipulate the entire autonomic nervous system. This is the neural wiring that supervises bodily functions beyond our conscious control. This complex system of nerves is organized into two opposing branches: the parasympathetic and the sympathetic. These nerves switch one another off and on, creating a system of checks and balances known as homeostasis. Acting in concert, these two complementary nerve systems enable us to make countless, nuanced internal responses to the ever-shifting demands of our internal and external environment.

For nursing mothers, the "environmental" demands are extraordinary. Their bodies must manage to produce thousands of extra calories per day to keep themselves and their babies alive. Also, their minds must find a way to block out external distractions and concentrate on their newborns if they are to glean the difference between a smile of joy or a grimace from gas. And above all, despite their overwhelming new responsibility, they must relax both body and mind so that their love and milk will flow.

Fortunately, oxytocin can act as the ultimate mother's helper. Oxytocin-producing nerves extend from the hypothalamus to connect directly to the vagal nerve—the largest branch of the parasympathetic nervous system. This nerve, when activated by oxytocin, fires up the gastric hormonal chemistry that digests food and turns it into glucose and fat that can be stored or fed to growing cells and babies. This caloric stockpiling gives mothers digestive powers that enable them to eke the most nutrition out of their diet so that they can stay strong and healthy even as their babies' hungry mouths demand more and more of them.

New mothers are under a great deal of stress, but they cannot afford to have exhaustion or worry eating up energy supplies. Oxytocin simultaneously provides this metabolic protection by quieting the sympathetic nerves and inhibiting the production of stress hormones that would normally trigger a catabolic or energy-burning response to such challenges. Oxytocin also gives mothers another powerful weapon against the ravages of stress: it slows their heart rate and lowers their blood pressure via its brain connections to key cardiac excitatory centers. The result is a coordinated antistress strategy that minimizes the wear and tear on a mother's body.

Lactation studies reveal that oxytocin does far more than redesign a mother's breast and gut functions; it also alters her mind. Kerstin Uvnas-Moberg, who is a gastroenterologist and a pharmacologist, has made a career of studying how oxytocin creates the hypermetabolism that supports lactation. While she was breast-feeding her fourth child, she began to wonder if this remarkable molecule might also be responsible for the sense of dreamy contentedness she felt while nursing.

Uvnas-Moberg returned to her lab at Sweden's Karolinska Institute to design a series of studies that would compare the mental and physical condition of breast-feeding and bottle-feeding mothers with the amount of oxytocin circulating in their blood. She found that in addition to having higher levels of oxytocin than nonnursing mothers, breast-feeding women feel less aggressive and less anxious, as well as less suspicious and less guilty. They also enjoy greater physical ease with fewer complaints of gut and somatic anxiety and experience less muscular tension than bottle-feeding mothers.

But not all nursing mothers are alike. Their scores on personality tests varied depending on their oxytocin levels. The mothers with the highest levels of oxytocin were the ones who experienced the most dramatic psychological, physiological, and behavioral changes. These were the mothers who reported feeling the most relaxed, most willing to make social contact, and most content to sit for hours feeding and studying their babies. Their survey answers also indicated that they felt the most attuned to their infant's needs. In another study, Uvnas-Moberg and her team also found a correlation between high levels of circulating oxytocin in breast-feeding mothers and the amount of time they spent fussing over and handling their infants. Uvnas-Moberg sees this as further evidence that the extra infusion of oxytocin these women received while breast-feeding enhanced their ability to refocus their metabolic, psychological, and behavioral priorities.

Psychiatrist Kathleen Light headed up a research team that supported Uvnas-Moberg's findings. Light compared oxytocin levels in breast-feeding and bottle-feeding mothers while the women completed a verbal stress test. They found that 50 percent of the breast-feeding women had stronger increases in oxytocin levels in response to the challenge while only 8 percent of the bottle-feeding mothers experienced an oxytocin surge. These oxytocin increases were accompanied by lowered blood pressure both before and after they made a speech about their life problems. The mothers with the strongest oxytocin response also reported feeling less angry and hostile, and more positive. They also had more interpersonal interactions than those who had produced less oxytocin.

This radical change of heart and mind that emerges toward the end of pregnancy and lingers through the first few months of a baby's life is something pediatrician and psychologist Donald Winnicott calls "primary maternal preoccupations." So intense is this newfound attraction that Winnicott describes it as "almost an illness." He observes that this compulsion to focus on her infant to the apparent exclusion of all else leads a mother to "heightened sensitivity and sense of responsibility."

Thanks to oxytocin, new mothers are in the mental and physical shape to stare at their babies for up to fourteen hours a day. This intense visual contact has the power to release more maternal oxytocin. The mere sight of her baby can cause a mother's milk to flow. (This same visual release of oxytocin and milk occurs when a cow sees the familiar silhouette of the dairy farmer in the barn door.)

The result is a cycle of oxytocin exchange that sharpens a mother's focus so that she is able to anticipate her baby's needs and learn the meaning of the sounds and signals in the infant's nonverbal repertoire. It is in this special oxytocin-enhanced mind-set that mothers and babies come to recognize and love each other. They "own" each other.

On closer examination, a mother's extremely convenient fixation on her baby is fueled by an intense sense of curiosity. Every little thing a baby does is magic—to a mother at least. This is nature's trick, being played to great advantage by oxytocin. Brain researchers Andreas Bartels and Semir Zeki used fMRI imaging to see what goes on in a mother's head when she's looking at her baby. In their study they didn't even have the mother look at her actual child but merely a photo of the baby. "Just watching" the image of her child was enough to trigger a powerful response in brain cells loaded with oxytocin that create pleasurable sensations and a deep sense of maternal commitment.

There seems to be a common chemical thread between Winnicott's primary maternal preoccupation, Panksepp's mere exposure, Shepard's tireless scrutiny, and Wilson's hunter's trance. All are visually based systems of attention. All are conducted with a slowed heart rate, a calmed mind, and quieted breath. All create a sense of heightened curiosity and connection that allow mothers—or hunters, scientists, and pet owners—to lose themselves in the minutiae of the moment and thrill to subtle changes that impatience and boredom would entirely miss. All are functions of brain systems heavily influenced by oxytocin.

Taken together, human and animal studies confirm that oxytocin is not just responsible for the dramatic physical effects that occur in labor and lactation but it also creates the relaxed physiological and psychological climate that keeps mothers and babies safe and strong while encouraging them to get to know and love each other. The attraction oxytocin inspires can be so intense that it leaves indelible bonds between mothers and their children.

From day one, all social mammals learn that the warmth and touch of another is good. The lesson is deliberately vague—more like one of Wilson's guiding emotions—than a hard-and-fast rule. This allows these first good impressions to resonate throughout the rest of a mammal's social life. The guiding emotions chemically forged while we are cared for as infants become a prerequisite to the formation of *all* of our social bonds. Long after we are weaned, oxytocin continues to keep us alert to the behaviors that invite friendly contact with others, and it helps us remember who our friends are.

Oxytocin is equally essential to the formation of the human-animal bond. As we will see in coming chapters, it flows through and between all mammals. We will also see why humans and animals can easily trigger this same chemistry of attachment in each other with a simple glance, word, or gesture.

Our new, improved understanding of this molecule's ability to create strong feelings of attraction, recognition, and commitment between mammals—and the fact that its effects can be released visually as well as through other sensory contact—strongly suggests that it was one of the subliminal forces that shaped the minds and hearts of our Ice Age ancestors.

If "just watching" the living world around us stirred this powerful socializing brain chemistry in our hearts and dawning minds, we could have stared our way into a growing sense of kinship between us and "them." This merging of identity would have bred an intimacy—a sort of primary maternal preoccupation—between humans and animals, plants, and places that would leave a vivid and lasting impression just beneath the surface of our psyches, an impression poised to reemerge in the thrill a child feels for a pet or the wonder that overwhelms us at the edge of a tidal pool.

Oxytocin could have ushered these ancient guiding emotions into the modern world because they have always made us feel better, think better, and behave better toward ourselves and others. It is this common oxytocin heritage that continues to promote feelings of familiarity and acceptance among and between mammals. Genuinely universal needs may indeed be hard to find, but for social animals, making social connections is essential. In oxytocin, we find the primary genetic ingredient of a general system of attachment that can support a phenomenon such as biophilia.

In the coming chapters we will see just how essential this transcendental affiliation with the natural world still is and how it continues its timeless business through the bonds it forges between us and our animals.

THREE

A Mind on Nature

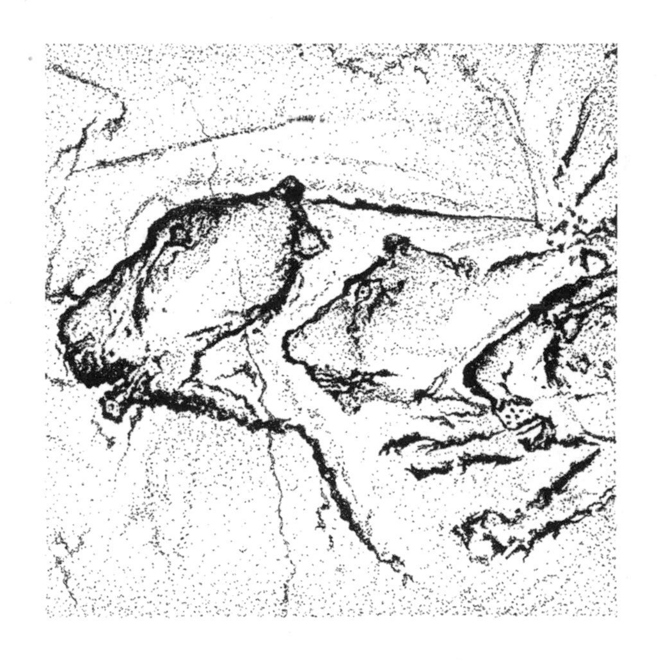

In our deep past, we were closer to animals in every way. We were surrounded by them, we thought more like them, and we even acted more like them than we do today. In *The Dawn of Human Culture,* archeologist Richard Klein says that until forty thousand years ago humans could still be considered "a relatively rare large mammal." And like our fellow mammals, we took our inspiration and cues from the forces of nature. Our world was immediate and much of what we knew about it we learned from watching.

How closely were we watching animals? Very closely, according to biologist Craig Packer and archeologist Jean Clottes. They made this assessment after looking at thousands of lion paintings that were drawn on cave walls in southern France thirty-two thousand years ago:

> Anyone who has spent time studying and photographing African lions will be stunned . . . For ancient artists to have made these observations, the lions must have been very relaxed in their presence. Modern-day African lions almost always flee from Masai pedestrians: tourists can approach within several yards only if they remain in their vehicles.

Only with such privileged proximity, Packer and Clottes say, could the artists have captured such accurate details of the lions' anatomy and behavior:

> Several pairs sit parallel gazing at the same distant object—behavior often observed in Africa . . . the postures are accurate and the facial expressions precise . . . Several of these portraits are so detailed that

> they depict the varied patterns of whisker spots, which biologists today use to identify individual lions in the field.

So they *knew* these animals—not just as a species but as individuals. These were neighbors, close neighbors.

This impressive detail and graphic skill tell us a lot about the animals we once couldn't take our eyes off. It tells us what those animals looked like and how they behaved. It tells us those animals were not terribly frightened of us. It also tells us some interesting things about the people who painted them. It tells us their eyes and visual memories were extremely sharp. It implies that there were people who were not scared to death of lions. It even tells us these artists were clever enough to see charcoal as more than the remnants of last night's fire. But the cave art does not tell us the one thing we want to know most. What were these painters thinking? What could they think? Did all that attention to detail and graphic skill mean the human brain had finally reached that level of awareness we call consciousness? Not necessarily, according to psychologist Julian Jaynes.

Jaynes made the case that many brain activities we consider advanced, such as different kinds of learning, concept making, problem solving, thinking, even reasoning, take place unconsciously. In fact, he explained, being conscious not only doesn't help these essential mental enterprises, it actually hinders their execution. For instance, Jaynes said, to play a piano, dance, run a race, or play tennis requires complex motor skills that may be learned with a degree of consciousness but cannot be performed successfully with it.

Psychologist Nicholas Humphrey has seen evidence that painting animals also belongs on that list. He says he has seen such talent flow from an artist who has neither language nor consciousness, a young autistic girl named Nadia. Nadia belongs to a rare class of autistics—the savants—who, despite their severe developmental disabilities, are capable of mastering extraordinary skills. Nadia's mysterious gift was drawing. By the age of three, she had shown no language or social awareness—two developmental deficits that severely limited her level of consciousness. Physically she was clumsy except in one striking way—she could draw animals and people from memory with an uncanny photographic accuracy.

In this and other ways Nadia had little in common with most of us, but in her strangeness Humphrey saw striking similarities to others, from another time. Her renderings of horses, cows, and elephants, Humphrey notes, are remarkably similar to the horses, aurochs, and mammoths drawn with similar skill thirty thousand years ago:

> in both cases, the graphic techniques by which this naturalism is achieved are very similar. Linear contour is used to model the body of the animals. Foreshortening and hidden-line occlusion are used to give perspective and depth. Animals are typically "snapped" as it were in active motion—prancing say, or bellowing. Liveliness is enhanced by doubling-up on some of the body contours. There is a preference for side-on views. Salient parts, such as faces and feet are emphasized—with the rest of the body sometimes being ignored.

Humphrey believes that Nadia's paintings suggest that cave art could have been produced by humans with limited consciousness and language.

Nadia's spontaneous artistic ability arose out of brain circuitry capable of producing a pathological version of tireless scrutiny known as extended selective attention. Autism expert Charlotte Modahl says that this extended selective attention arises from a delayed ability to shift attention from one set of incoming environmental stimuli to another. This incoming stimulus gets played over and over by the feature-analytic cortex of the brain where it intensifies the perceptions of that stimulus, resulting, Modahl says, in "supernormal high-fidelity visual or auditory memory."

The hi-fi savant mind can also show a knack for numbers or grasp the passage of time in ways the normal, conscious brain cannot. Calendar-calculating savants may not be able to do simple math, but they can instantly tell you that April 11 will fall on a Sunday in 2190. Interestingly, such prodigious feats of calendar calculating may not have been unknown to the watchful prehistoric humans who could also predict exactly where the dawning rays of the summer solstice would fall.

A very important difference, of course, between the first artists and Nadia is that Nadia's skill comes from an interest that is considered abnormal. The cave painters' interest in animals encouraged an attention span that was appropriate to their environment. Their talent may have been exceptional, but their obsession with their subjects was normal. Animals dominated all aspects of their lives. In fact, it would have been almost impossible for them to turn their gaze away from these omnipresent creatures. Their art was created in a world order very different from ours, when the ability to hold such stimuli as animals in tight, extended focus proved to be highly adaptive.

Archeologist Stephen Mithen envisions the early human mind like this:

> In my view, the early human mind may have been similar to a Romanesque cathedral. These are characterized by having several chapels separated from each other by thick walls and low vaults, so that the sounds of the services of one chapel are almost inaudible elsewhere in the cathedral. The intelligence that early humans used to make stone tools, or to understand animal behavior or the social world, may have been "trapped" in parts of the mind like these chapels unable to be heard by the rest of the mind.

Might animals "trapped" in these ancient neural chapels have been subjected to a kind of hyperscrutiny that imprinted their anatomy and behavior into our preconscious brains? If such vital information can be gleaned in this way, our brains seemingly would have retained these valuable ways of knowing without knowing.

Savants are completely unaware that they are performing awesome mental feats. In this cluelessness they differ from the rest of us merely by degree. Many geniuses have testified to the nonconscious nature of their most profound revelations. And we've all had those "aha!" moments when realizations pop effortlessly into our heads as if from thin air. If our brains sometimes seem to be smarter than we are, it is because, Jaynes says, "What you can consciously recall is a thimbleful to the huge oceans of your actual knowledge."

In an attempt to understand how accessible nonconscious learning is to a modern, nonautistic mind, a group of undergraduates were asked to locate a target number appearing at different locations in a series of matrixes projected onto a screen. Unknown to the students, the placement of the target number was determined by a formula that was revealed in some of the simpler, earlier patterns that flashed before them. By measuring how long it took the students to spot the target number as the matrixes became more complex, the examiners could tell when the student had figured out the formula being used.

After more than twenty thousand exposures (a kind of "extended selective attention"?) the students' response time quickened, indicating they were using the code to predict the target's placement. Yet none of the participants reported noticing the formula at the beginning of the test or being aware they were using it to locate the target number as the test continued. Tireless scrutiny, not consciousness, supplied this insight.

Tens of thousands of years of staring at animal targets or being stared at by animals also left our early ancestors masters at the mental art of "thin slicing." In his book *Blink*, Malcolm Gladwell describes thin slicing as an ancient right-brained way of knowing that provides us with instant and effortless assessments of what we see, the second we see it. These assessments are not open for discussion. This is strictly preverbal know-how. Any attempt at deliberation slows it down and distorts it hopelessly. In seconds, our brain decides if we are comfortable with another person or animal based on a vast store of visual and behavioral patterns we've logged away for just this purpose.

Every second of every day, we are designing our lives based on information we have no idea we are receiving. Fortunately, this silent system is damn good. Even for the most oblivious, it works with surprising accuracy. But it could be better, and it is for those who spend more time just watching. The eye can be retrained to magnify the bit of information coming at us through these ancient visual pathways. And then there are those who never forgot the art of silent seeing.

The accuracy of what we blindly see in the blink of an eye is only as good as the staring that preceded it. Fortunately, our ancestors had the powers of concentration to amass the sort of intelligence that made survival possible. We all retain this ability in varying degrees, but to see a visual brain perform at the top of it game, we must go to an island in the Pacific that civilization and technology forgot.

Through an accident of geography, Papua New Guinea missed out on the Neolithic revolution that gave us farming and farm animals as well as the industrial revolution and the modern world. It's not that time stopped on New Guinea—the natives developed language and made art, but they also stuck with those same old stone tools that had built their world for millions of years and they never took to farming. Despite their conversations and decorations, they remained Stone Age hunters and gatherers for another seven thousand years, which is why Papua New Guinea offers us the best chance we will ever have to "see" the world as our undomesticated, uncivilized ancestors once did.

For twenty years evolutionary biologist Jared Diamond sought to understand how Stone Age New Guineans see their world. In the end, he realized, no matter how hard or long he stared, he couldn't even look at a bird the way they do. He was particularly humbled by his efforts to identify a couple of species of warblers he'd collected on a visit to the Foré tribe. Even holding them dead in his hand—with the entire ornithological collection of the American Museum of Natural History for comparison—he took weeks to detect their subtle distinctions. His sense of accomplishment was checked when he discovered that the Foré were well aware that the birds belong to two different species—which is why they had separate names for them. "To make matters more embarrassing, the Foré distinguished the *mabisena* and *pasagekiyabi* in the field without binoculars, in silhouette, when the birds were 10 meters away." In the jungles of New Guinea, Jared Diamond was the primitive.

On another island, Diamond recorded one man's bird knowledge: "For every one of Kulambangra's eighty resident bird species, Teu dictated to me an account consisting of its name in the Kulambangra language, its song, preferred habitat, abundance, size of the group in which it usually foraged, diet, nest construction, clutch size, breeding season, seasonal altitudinal movements, and frequency and group size for over-water dispersal." This fantastic natural database is maintained mentally by the island's Kalam people. With language they can name each creature and share this information. But it is watching and listening—*pre*verbal, nonconscious, archaic mental skills—that enable them to recognize more than fourteen hundred wild animal and plant species. It makes you wonder what a jungle looks like when you know every plant and animal in it.

The Foré retained their Stone Age priorities for seven thousand years longer than we did. They've entered the twenty-first century with a mind that still caters to these primal purposes. They tap into the wiring that gave Nadia her acute awareness of the animal form and the calendar savant's ability to absorb and instantly retrieve vast amounts of information about patterns in the natural world. Clearly, there is more to the mind than we can "see." The modern Foré mind retains outcroppings of the archaic mental layers surmised by Jaynes. While the Forés' extraordinary talents of observation may appear savant-like, they are the result of adaptation, not malfunction. This is your mind on nature—a mind that specializes in visual intimacy with the life around it.

FOUR

The Rules of Engagement

As possible as it is to feel an affinity with all living things, we humans tend to fall for creatures that are warm and fuzzy. E. O. Wilson, an ant enthusiast, acknowledged that while we are kin to all organisms, most of us have a sort of biophilia bias for members of our own immediate family—the mammals.

We mammals got our good name from secretory tissues we have called mammary glands that can provide complete nourishment for our offspring. In the eighteenth century, Linnaeus, the father of taxonomy, chose this anatomical feature to distinguish us because of its essential role in the survival of our species and the fact that it separates us from our reptilian and avian cousins. By selecting mammary glands and their milk-giving function as our defining feature, Linnaeus zeroed in on the essence of all things mammalian—including our social behavior.

Lactation is one of nature's masterpieces—brilliant, elegant, and expensive. Our species thrives on our ability to allow our young to suck away our vital fluids. This is the cost of doing mammalian business. To meet that cost, as we have seen, the biology of lactation also comes equipped with the capacity to produce a complementary metabolism and nervous system that can nourish both mother and child. Still, all the cleverly designed brain and body parts that go into making and delivering breast milk are useless if a mother won't lift her baby to her breast or the baby won't suckle. And so it makes perfect sense that the system which delivers mother's milk also provides the magnet that draws us to each other.

All manner of social interaction is actually a variation on the three phases—hunger, eating, satiety—of our infantile feeding experience. For the rest of our lives we will hunger for touch, have sexual appetites, and sleep like a baby afterward. This likeness was obvious to Kerstin Uvnas-Moberg, who has explored oxytocin's hidden powers over lactation physiology and social attachment. She describes their inescapable connectivity this way:

> All affiliate behavior takes place in three stages. When an encounter is perceived as non-threatening, oxytocin will encourage approach. That's followed by interaction, which can range from sexual behavior and lactation to all kinds of less specific interactive behaviors involving touch or other sensory stimuli.
>
> The oxytocin released during these encounters coordinates a myriad of physiological reactions such as lower pulse rate, blood pressure, and stress hormone levels. At the same time oxytocin promotes the restorative bodily functions like energy storage and growth. The energy conserved by these reactions produces the final stage of positive social interaction—relaxation.

Whether it's good mothering or a good meal, Uvnas-Moberg says, the hormonal rules of engagement are the same. This is the essential mammalian social etiquette learned, quite literally, at our mother's breast, and the primary author of this social code is oxytocin.

~

Oxytocin appears to be an elaboration of more primitive chemicals that have promoted growth and reproduction since the appearance of single-celled animals. Growth was, in fact, the first reproductive act. Growing is how single-celled animals "do it." They expand till they divide. Eventually single-celled animals came together into multicellular life forms capable of more complex functions and behaviors. Perhaps some ancient ancestor of oxytocin had a role in creating this original and brilliant bonding strategy.

Cooperation and complexity gave birth to reptiles, birds, and fish—all creatures equipped with oxytocin-like agents that promote growth, instigate mating, and stimulate the muscle contractions necessary for things like egg laying. Even nonmammalian species have precursor versions of oxytocin that may account for their surprising social inclinations. Certain microscopic roundworms, for instance, seem to prefer eating in the company of others. Fish can be highly social, swimming in schools and caring for their young. Fishermen have even witnessed fish valiantly attempting to rescue "classmates" that have taken the hook.

Nature is like a farmer who keeps every truck and tractor he ever owned because he may someday need a part. We mammals are the beneficiaries of nature's compulsive hoarding and tinkering. After 200 million years of kicking old tires, nature cobbled together nine amino acids to form a better gadget—oxytocin.

Oxytocin seems to remember every evolutionary lesson it learned in earlier incarnations. It still knows how to divide cells and feed them. It continues to carry out all sorts of other basic functions like regulating body heat and fluid retention. It supports sexual reproduction and the distinctly mammalian chores of delivering babies and breast milk. Oxytocin not only serves our first meal, it continues to regulate our eating habits for the rest of our lives. Oxytocin can both decrease and increase our appetites. It can encourage us to seek out food or sense a full belly and tell the brain that we are sated. Oxytocin even gives us that sleepy feeling that overtakes us after a big meal. Since positive social interaction is essential for mammalian growth and reproduction, it too falls under oxytocin's jurisdiction.

Oxytocin could never accomplish all this on its own. Although oxytocin is one of the most powerful neurochemicals our bodies make, it has less than twenty minutes in the bloodstream and no more than one hour in the central nervous system to act before it is broken down and absorbed by the body. Fortunately, in the first few days of life our baby brain responds to our mother's warmth, touch, and affection with a bloom of oxytocin receptors throughout the many brain areas essential to social recognition, remembrance, and reward. Oxytocin experts believe it may be this early, saturating presence of oxytocin throughout our social brain network that sensitizes these cell bodies so that throughout the rest of our lives, even a brief encounter with oxytocin will have a potent effect on a host of other chemicals in the brain that can help to fulfill oxytocin's essential social agenda.

To begin the business of bringing mammals together, oxytocin must first overcome the social reticence of the amygdala, also known as the "fear center" of the brain. When it is presented with ambiguous sensory or emotional information, the amygdala can instantly activate the defensive neural network in our brain that creates the decidedly antisocial fight/flight response. Fight/flight has gotten us out of trouble for millions of years, but for social mammals, it is equally important to be able to welcome the approach of others. So we must be able to give this reflex a rest, and that is another thing oxytocin has evolved to do.

Oxytocin is able to calm the paranoid tendencies of the amygdala by activating cells in the center of this nerve cluster that release one of the body's natural tranquilizers, a neurotransmitter called GABA. Its calming influence prevents the amygdala from automatically perceiving new or ambiguous faces, places, or ideas as threatening. It even improves the amygdala's ability to recognize the subtle signs we send through posture, voice, and facial expressions that signal friendly intentions. With GABA's assistance, oxytocin helps the amygdala sharpen our social perceptions and remember those favorable first impressions later. The result is an amygdala chemically tuned to accept social approach.

Oxytocin also holds a calming sway over the rest of the fight/flight defensive network. For instance, if the amygdala excites the sympathetic nerves, they will cause the arousal centers of the brain to release noradrenalin, which gets our hearts pumping blood to our muscles and makes us feel anxious. Oxytocin can block this effect by dramatically increasing the number of noradrenalin's inhibitory receptors that act as off switches. In this way, oxytocin not only prevents a rise in heart rate or blood pressure but actually causes both to be lowered, producing a sense of well-being.

The other key mechanism that the amygdala relies on to mount a defense is a Rube Goldberg–like chemical cascade called the HPA stress axis. The H is for the hypothalamus, which releases CRH (corticotrophin-releasing hormone) and vasopressin. In the last chapter we saw vasopressin instilling a sense of fearlessness and aggression in male prairie voles in the social cause of guarding their mates and babies. However, vasopressin is better known as a stress hormone that raises blood pressure and conserves bodily fluids. It is also a key instigator of fight/flight.

Once vasopressin and CRH cells in the hypothalamus are activated, they will trigger the P (pituitary) gland's release of ACTH (adrenocorticotrophic hormone). When this blood-borne stress chemical reaches the A (adrenal) glands, it prompts them to release the stress hormones adrenalin and "cort" (cortisol in primates and corticosterone in most rodents). Adrenalin and cort increase blood pressure. Cort also requisitions stored-up calories to provide the bone and muscle power to run away or stand and fight.

This is the chemistry that overcomes two male rats when placed in a brightly lit, open space. Male rats are not social animals to begin with, and they feel very vulnerable in such exposed circumstances. They react in classic fight/flight behavior by either rearing up in defense or clinging to the wall in an attempt to disappear. A dose of oxytocin, however, not only lowers their HPA chemistry but causes the rats to venture into the center of the enclosure, where they sniff each other with friendly curiosity. This willingness to interact signals that oxytocin has not only overcome the brain networks that prevent social approach but has now engaged the neural circuitry that will actively encourage the second phase of the social contract: interaction.

Now oxytocin calls on the dopamine-producing brain cells to provide the actual physical push we all need to take those first steps in seeking out potentially satisfying social encounters. (People with Parkinson's disease have low levels of dopamine, which causes tremors, jerky motions, and in extreme cases, an inability to move forward.) Oxytocin injections can release dopamine, while an increase in dopamine can trigger a rise in oxytocin. This mutual feedback system provides the chemical nudge we need to engage in social contact.

The receptive state of mind created during the approach phase promotes all kinds of positive interactions from fruitful conversation to casual touch to sexual intimacy. When the interactions involve welcome touch, pressure-sensitive nerves in our skin convey signals to the oxytocin-producing nerves in the brain that lower heart rate, blood pressure, and our stress chemistry. They also cause a rise in the feel-good chemicals, dopamine, serotonin, and beta endorphin, that mark the occasion as enjoyable. The end result is a social encounter that produces a rewarding sense of relaxation and satiety. This also creates a fond memory that, in the future, will whet our appetite for more.

This state of calm/connect is far more than a default condition we shift back into when we are not in some version of the fight/flight mode. Researchers are just recognizing how this dynamic chemistry works and why it is essential to the reproduction and well-being of all social mammals. In the coming chapters we will see how this oxytocin-orchestrated brain network also came to calm and connect humans and animals, producing social bonds that would prove to be more satisfying and more enduring than we ever could have imagined.

FIVE

Brave New World

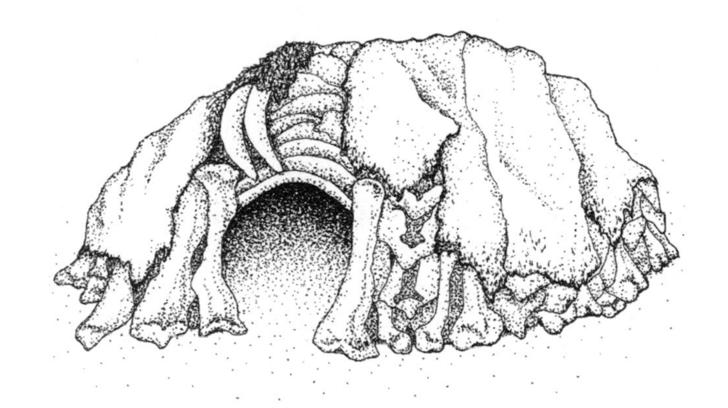

We are the top predators of our world, and the lives of stockbrokers and even vegetarians are defined by this reality. Our Ice Age ancestors, however, were far from alpha, and they knew it. They lived in a climate of fear, outnumbered and overpowered by many other animals. Despite ridiculous odds, something changed; they found themselves rising up to enter into close, deadly combat with the beasts that had cowed them for so long. To get a sense of the Ice Age hunter's shaky self-esteem, we must try to understand his primitive consciousness, tailored by a natural order very different from our own.

For a million years, we drifted along the margins of vast roving herds of animals like the ones that defied Buffalo Jones's reckoning. We will never see such a sight, but to our ancestors it was ordinary. This was what they knew as big game. They absorbed these spectacles, like the Foré, by visually breaking them down into salient details to be stored in the nonconscious part of their brain to be retrieved in an instant without debate. How they quieted their pounding hearts was another matter.

In what we now call the Ukraine, men dared to take on behemoths like the wooly mammoth. As early as twenty-seven thousand years ago they set up hunting camps on the edge of the mammoth's migratory path. And every year, for thousands of years, they would go there to kill or be killed.

Their shelters were monuments to their victory over great beasts and even greater fear. From the skeletons of the giant animals that fell before them, they sculpted something to call home. Here, under vaults of great arched ribs mounted atop stacks of giant mandibles, skulls, hips, and scapulae—

under twenty-three tons of bone—they wrapped themselves in the hides of their kills and dreamt dreams we will never have. These first ivory towers testify that the old order had been turned inside out. A few wildly ambitious humans, armed only with sharp spears and a growing sense of camaraderie and bravado, faced down giant enemies and bested them. Good eyes got us just so far with animals, but great nerve would take us the rest of the way. This chapter is about how we found that nerve—or nerves—that made us feel brave enough to approach the beasts we had just watched for so long.

For millions of years hominids were hunted—an evolutionary reality that dictated everything about the way their bodies and minds worked. Creatures that are preyed upon are quite sensibly fearful of strange sounds, touches, smells, or sights. For them, novel stimuli are instantly interpreted as threatening and trigger the full-blown activation of the chemistry of the fight/flight. The first great enemy that early humans had to conquer on their ascent from prey to predator was overwhelming fear. Literally, they had to change their minds—and bodies—so they could think and act differently when confronted by deadly animals. We know that about 100,000 years ago they managed to make this monumental attitude adjustment. But how? And why?

The first clue lies in the timing. Human behavior radically shifted in the Pleistocene, or the Ice Age. Every hundred thousand years or so, glaciers smothered Europe and forced humans and animals into close quarters. All living beings were forced to compete for the last of the water and the first of the new green. Only the most adventurous and curious animals would find these times interesting. Some were more curious than others, and as we will see, they inherited the earth.

The evidence that between 100,000 and 50,000 years ago our ancestors were facing this world with both a boost in brain power and nerve can be found in the weapons, the fires, the butchered carcasses of large animals, and in those amazing hunting shelters they left behind. All this tells us they had developed a new temperament. A closer look at the nature of that personality change suggests this mind-body shift took place in the bosom of Ice Age mothers.

About 1.5 million years ago, the human brain began to balloon. For Ice Age mothers this presented an obstetrical crisis. The pelvis that supported their upright posture and mobility was too narrow to accommodate the growing fetal skull. As anthropologist Meredith Small explains, "Hominid babies now had to twist and bend to pass through the birth canal, and more important, birth had to be triggered before the skull grew too big."

The Ice Age infants that did survive came into the world barely viable. These babies, like ours today, were born with a large brain only 25 percent "cooked." Compared with a rhesus monkey, whose birth brain is 75 percent wired, our babies are woefully premature, requiring a great deal more baby-sitting than the newborns of our closest ape relatives.

It became imperative to create a safe and nourishing environment outside the womb in which babies could continue growing smarter and stronger. Ice Age mothers were the first to cope with this new level of intensified parental obligation. Fortunately the oxytocin that helped to eject these underdeveloped babies and stimulate their mother's milk also enabled the entire clan to feel the urge to hold them close for as long as necessary.

The emotional transformation oxytocin can deliver is just being appreciated. It is intense enough to turn carefree young women into the preoccupied, even obsessive parents D. W. Winnicott describes. It can also inspire virgin females to take an interest in another's offspring, which would have come in handy once we started having babies that were more than one mother could handle.

Anthropologist Sarah Blaffer Hrdy says that being a good mother was not enough in an age of ice. Mothers with such fragile babies needed more hands and more help. Their ability to recruit allies became a matter of life and death. As Kerstin Uvnas-Moberg has found, even today, the oxytocin pulsing through a new mother makes her inclined to solicit help from others. In her personality studies of nursing women she also found that the more oxytocin they produced, the higher they scored for a personality trait called "social desirability." These oxytocin-enhanced mothers were also more "prone to please and obey during this period of life." If a surge of oxytocin brought out a similar personality shift in Ice Age mothers, it may have inspired their clan members to be more supportive and protective of them and their babies.

Being socially desirable may have been just as essential to surviving Ice Age childbirth. Anthropologist Wendy Trevathan thinks that the increased pain and difficulty of Ice Age labor may have led human mothers to be the first primates to have to ask for help delivering their babies. A female kin or clan member present during labor could have offered mother and baby critical assistance and support during this emotionally and hormonally charged event. These first midwives may have come away from the experience touched by the same bonding chemistry that turns even the most nonmaternal rodents into doting alloparents—a transformation that could have been critical to the survival of infants whose mothers died during labor.

Hrdy says that helpless Ice Age babies and toddlers also needed to be agreeable and pleasing to compete with a new sibling for their mother's attention or enlist the help of others in the group. This instinct for solicitation kicks in right after birth and is initiated through eye contact. Even infants only minutes old will lock eyes with adults and imitate their facial movements. This earliest form of flattery gets them everywhere and is made possible by the mirror neurons that allow tiny babies to imitate automatically the facial expressions they see another person doing.

If this charm offensive doesn't work, there's always the distress call. From the minute we're born, oxytocin also encourages us to communicate vocally. A simple defect in a gene for the oxytocin receptor causes mice pups to emit fewer ultrasonic distress vocalizations when they are socially isolated—a failure to communicate that can result in their demise.

Oxytocin also helps to make sure that our earliest cries are heard. As Hrdy explains, "all apes are attentive to the sights and sounds of newborns, and many (especially females) find newborns magnetically attractive regardless of whether they are related to them." Virgin California mice, who are ordinarily excellent baby-sitters, ignore the distress calls of mice pups—and even fear the babies—if they lack the gene to make oxytocin. Injecting them with oxytocin quiets their anxiety and restores their concern for the offspring of others.

Researchers from McGill and Emory universities were the first to notice the profound effect that maternal attention can have on the temperament and physiology of offspring. In 1998, they found that a simple increase in "arched-back" nursing (the fully engaged posture that attentive nursing rats assume) and pup licking could make rat offspring less fearful of novelty and more willing to explore the "open field" (or center) of their enclosure instead of clinging to its walls. They traced this new boldness to a reduced activity in the amygdala and the noradrenalin brain centers. They also showed that these pups had a suppressed HPA stress axis. Their groundbreaking findings appeared in a paper entitled "Maternal Care During Infancy Regulates the Development of Neural Systems Mediating the Expression of Fearfulness in Rats." Over the next decade, this team, lead by Michael Meaney, conducted additional studies showing that these antistress effects were permanent, instilling offspring with a confidence and curiosity that would keep them calm in the face of challenges throughout their lives.

The next revelation to come out of Meaney's lab was the discovery that the female pups who received high-quality maternal attention went on to be more maternal themselves. When they had pups, they cared for them in the same attentive way. Their pups were also less fearful and more adventurous. The third generation of daughters also inherited this lick-on effect. They proved to be better mothers of calmer babies, showing that early life experiences can permanently modulate the central nervous system as well as create a nongenetic pathway in which these physiological and behavioral traits can be passed on for generations.

Of course, any discussion of maternal behavior is by default a discussion of oxytocin. Michael Meaney and colleagues Frances Champagne and Darlene Francis showed that decreased licking by a mother rat weakened the oxytocin receptors in the fear circuitry of the offspring's brain, while increased maternal care developed greater densities of oxytocin receptors in brain regions known to regulate maternal behavior, fear, and stress. The team also found that male rat pups were affected differently by the added maternal attention than females. Young males left the nest with an increase in vasopressin receptors, which made them bolder.

Cort Pedersen found that giving oxytocin injections to new rat mothers makes them lick their babies more. He and colleague Maria Boccia also demonstrated that oxytocin causes rat mothers to nurse in the arched-back posture. When they chemically blocked oxytocin, they saw the mothers spent far less time in the arched-back nursing stance. They concluded that "oxytocin tilts the balance of oral grooming by dams away from themselves and towards pups." This oxytocin shift from self-interest to maternal concern proved to be just the ticket for creating generations of offspring better able to cope with stress and emerging opportunities.

Another way to get a mother lab rat to intensely care for her young is to remove her litter briefly. When the pups are returned, she vigorously grooms them and huddles over them. Perhaps Ice Age mothers, while grateful for a break, also smothered their babies with oxytocin-releasing attention when they were reunited.

Human Ice Age babies could have flourished in similar ways if they received extra "licking." The added boost to the oxytocin system would have had a calming and socially sensitizing effect that could help explain how and why we became the smartest animals on earth. "Individuals that could more accurately interpret events and skillfully maneuver in complicated and fluid social environments," Pedersen explains, "were more reproductively successful." This would have ensured that these oxytocin traits would be selected for, ushering in a new era of higher social intelligence.

Our deepening understanding of the biology of parental and alloparental behavior appears to support Hrdy's theory that a rise in Ice Age babyphilia could have been powerful enough to redefine our sense of social obligation. We now have neurohormonal reasons why baby-sitters would have become more available and perhaps more competent. It even explains how highly protective mothers could have become convinced that it was safe to pass their babies off to the care of these nurturing females.

Many hands may have made lighter maternal work in the Ice Age and greased the oxytocin gears that made us more interested in than afraid of each other and the world around us. A powerful attraction to and empathy for babies is evidence of a mind focused by oxytocin—and so is the growing sense of trust that spread through mothers and others during the Ice Age.

~

Compelling evidence that an early infusion of oxytocin could have expanded our Ice Age social imaginations comes from two recent studies conducted by a new breed of researchers known as "neuroeconomists." Paul Zak is a pioneer in this field. He teaches economics at Claremont Graduate University and neurology at Loma Linda University. This unusual combination of expertise led Zak to question the role neurochemistry may play in encouraging trustworthy behavior—the cornerstone of social and economic development.

Zak's recent studies have examined oxytocin's influence in creating the sense of trust necessary to the exchange of money. He measured oxytocin levels in students who participated in a classic economic experiment called a "trust game." This game is played with two groups that are each given ten dollars. Half of the subjects can transfer some or all of their money to the other half. Those on the receiving end will have that amount tripled in their account. They can then return some or none of their enhanced kitty. All subjects know the rules and all are anonymous.

Zak says that while all the risk falls squarely on the first group, they still tend to be generous in their transfers. The receiving group historically sends back at least some of the windfall, even though they have no obligation to do so. Zak suspected that the generosity of the first group was seen by the recipients as a show of trust. Zak wondered if that "trust signal" might not trigger a rise of oxytocin in the beneficiaries, making them feel a sense of connection and obligation that increased the amount they returned.

When he analyzed the blood samples from the study group, he found that those who had received generous transfers had a spike in their oxytocin levels. The more money they received, the more they returned. Reciprocity is how economists measure trustworthiness. Zak's study suggests that acts of trust can inspire oxytocin and trustworthiness in others.

Zak also participated in a similar study conducted by Swiss neuroeconomists. This group led by Ernst Fehr, a professor at the University of Zurich, wanted to see whether oxytocin influences how much the "investor" group decides to send to the "trustees." In their game the rules were the same, but half the subjects were given an oxytocin inhalant and half were given a placebo. (In Europe, oxytocin nasal sprays are used to stimulate labor and lactation. Inhaled oxytocin has also been shown to permeate the brain, where it can affect behavior.)

Fehr found that those who inhaled oxytocin invested 17 percent more with their trustees than the placebo investors did. And twice as many oxytocin investors transferred the entire sum to trustees compared with the nonoxytocin investors. However, when investors were told that they would be playing the same game, but this time their return would be calculated by a computer programmed to respond exactly as a human trustee would, even the investors under the influence of oxytocin held on to their money. Fehr concluded from this machine versus human distinction that oxytocin increases a person's willingness to take not just any kind of risk, but *social* risk.

Studies like these help to explain why it's plausible that an early increase in oxytocin could have sent a wave of "trust signals" around campfires throughout the Pleistocene world. The effect would have created tribes of less paranoid mates and hunters, more reliable, enthusiastic baby-sitters, and mothers who were more trusting of them. It may seem ridiculous to lay all that responsibility on one tiny molecule, but there is an evolutionary logic to it. Interesting times call for interesting—and interested—people. Mothers and helpers with open hearts and minds could have populated the volatile Paleolithic landscape with men and women with the increased social intelligence that was perfectly suited to the times.

Smarter, more sensitive and patient cavemen would have felt a growing sense of recognition and commitment to the mother of their child and perhaps to the child as well. Recent research shows that stable, supportive relationships enhance the production of oxytocin in both men and women. But the emerging paternal personality of Ice Age fathers could also have received a big assist from vasopressin. In monogamous rodents like the prairie vole and California mouse, attentive fathers can produce a lick-on effect of their own that predisposes their sons to be better fathers.

In male prairie voles, as we saw in Chapter 2, their monogamous and paternal devotion is linked to the length of the regulatory allele for the V1a vasopressin receptor. Fathers with long alleles lick their sons more, increasing the production of vasopressin receptors in their social brain networks. These well-fathered sons then go on to do the same for their sons, and so on.

In humans, the V1a vasopressin receptor gene also has a stretch of repetitive DNA similar to the one that fine-tunes the level of sociability in male prairie voles. In us, the length of this allele may be linked to our capacity for altruism. Two hundred and three male and female students from the Psychology Department of Hebrew University in Jerusalem played a one-sided economics game designed to measure altruism versus self-interest. In this "dictator game," one group of students gets money and transfers whatever amount they decide not to keep for themselves. Game over. Researchers found that the students with the longest V1aR allele gave away the most money, leading them to conclude that "the same gene contributing to social bonding in lower animals also appears to operate similarly in human behavior suggest[ing] a common evolutionary mechanism."

Another study supports a link between genetic modulation of the V1a vasopressin receptor and the human capacity for pair bonding. Men with two copies of a particular V1aR allele were more likely to be unmarried and if married, reported more marital strife than men without it.

During the Ice Age, male-female relationships became more defined and secure, and babies weren't the only ones to benefit from that. The need for males to defend their mates or constantly compete for sexual favors gradually abated, leaving room to explore new relationships—further evidence of oxytocin's expanding influence. Ice Age humans not only became more social, they became braver. Well-tended Pleistocene babies found themselves guided by a more oxytocin-friendly central nervous system that shrugged off the suspicions and anxiety of the old prey mentality. A calmer, more curious male would have become more willing to trust his fellow cavemen to cover his back—a necessary prerequisite in the neuroeconomics of big-game hunting.

Well-raised cavemen came together in newly cooperative, noncompetitive ways. United, they stood taller and stronger against the great beasts surrounding them. Their deeper sense of trust in each other emboldened them to approach and even enter the attack zone of predatory beasts. Kerstin Uvnas-Moberg discusses the different emboldening qualities of oxytocin and vasopressin in her book, *The Oxytocin Factor*. "Vasopressin instills courage by making the individual feel aggressive and fearless. . . . Oxytocin instead fosters courage by diminishing the feeling of danger and conveying a sense there is less to be afraid of." Both brands of courage were needed to take down the big game. There is no doubt our ancestors somehow found that courage in themselves and in each other.

We tend to think of hunting as one of the first inspirations for increased social cooperation, but it may be the other way around. Parental love is the first and most essential expression of social cooperation. An increase in parental attention could have created the neurohormonal atmosphere that helped us grow less wary of each other and the animals around us. In bands of newly minted "brothers," we bolstered each other's courage until our ancient, submissive self-image gave way to the bold ego of the hunter.

In big-game hunting, the ancient social contract was supersized. The heroics and the rewards exemplify approach, interaction, and relaxation like never before. The goodwill created by a more altruistic division of the spoils would have further united and satisfied the community. Top that off with the wave of oxytocin released by the warmth of the campfire and a belly full of protein and fat, and you've passed relaxation and gone into stupor.

As human carnivores learned to appreciate the advantages of expanded social alliances and the taste of fresh-killed meat, a more relaxed and adventurous attitude encouraged humans to creep closer to animals. Most used their new access to kill them, while some may have just wanted to get a better look. They were the gifted and talented Ice Age children who got close enough to capture the likeness of their prey in paintings on cave walls all over Europe.

As we saw, these earliest depictions of animals are so accurate that even tens of thousands of years later, we too can know the animals our ancestors knew. They are the product of the kind of intense devotion that leaves a neurobiological mark on social mammals. Ancient folk tales from the last Ice Age cultures say that those who drew animals became them. Oxytocin and vasopressin must have played an important role in the dawning recognition of animal-as-self that gave rise to shamanistic mergings and the frenzy of creativity that was stirring around campfires.

In the middle of the Ice Age, the human heart was melting. Cave artists felt it first. The beast had become more than dinner; it had become Muse. Over and over, tens of thousands of times, cave painters made horses and lions and aurochs, and they made them do whatever they wanted. Cave artists were the first humans to gain control over wild beasts. They alone could decide where an animal would be and what it would do. Perhaps in these caves, in this art, humans first toyed with the notion of finding another way with animals.

Their sharp eyes must have noticed that the animals they watched so tirelessly were acting differently too. The long, stressful Ice Age was no less formative for the other mammals that survived it. So it may be that the same epigenetic forces that made Paleolithic people more sociable were having a similar effect on another social carnivore—the wolf.

If a new strain of daring and charming wolves followed their hearts and noses into the world of the cooking carnivores, they would have entered a society neurobiologically prepared to bond with them. If their curious pups followed them, they would have been gazed upon by humans whose nurturing instincts were at an all-time high. Oxytocin and vasopressin made us so parental we may have begun to see animal babies as our own. We found ourselves lavishing care on babies and wolf puppies—a pastime that, as we will see, had dramatic neurobiological consequences.

In the end, we and our wolves emerged from the Ice Age more optimistic, gregarious, content, and cooperative. We emerged as friends. In the coming chapters we will see how biology formed this human-wolf bond and made domestication possible.

SIX

Good Dog

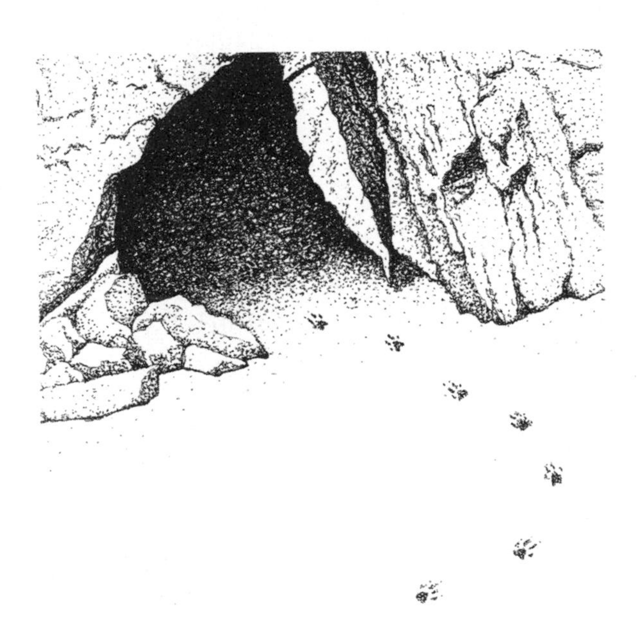

Around 125,000 years ago, some humans were living inside a cave in southern France. At the entry to each living area, the residents placed wolf skulls facing out. Did they represent a warning, a gesture of admiration, or bragging rights? We can't know what the wolf skulls meant to those people, but we can be sure that the careful and prominent placement of those skulls was the thin end of the domestication wedge. Over the next hundred thousand years the descendants of those animals would slowly worm their way deeper into homes and hearts. Wolves would teach us a new idea called "tame," and it would change the world.

In Chapter 1, we saw that the earliest humans had to face the challenges of living in a world best suited to social carnivores. Harry Jerison says this was no easy trick for a primate. "The model of species adapted for such a niche is the well-studied timber wolf, a proper social carnivore, with the proper profile of morphological, neural, and behavioral adaptations for life in this niche." Wolves may have also been the carnivorous species our ancestors modeled themselves after.

For a million and a half years, humans watched wolves and ate their leftovers. Then, about 400,000 years ago, they began to hunt for themselves. They fanned out into wolf territory and stalked the same game. Humans may have even learned a thing or two about driving herds and rounding up their prey from these expert social hunters. In *A History of Domesticated Animals*, archeologist Frederick E. Zeuner includes an aerial photo of wolves driving a herd of caribou to make his point that "the wolf knew how to round up ruminants long before man thought of doing so."

Paleolithic humans, as the cave paintings prove, were brilliant students of animal behavior. The first hunters would have long observed the wolves' cooperative strategies for stalking herds and isolating prey. Wolves are watchful too. They would have spotted our familiar hunting tactics and increasing success. At some point wolves must have stopped watching and approached the new predators. Rather than compete with this new hungry hunter, wolves may have opted to play their own version of the trust game. They would join these humans in their chase, combining their superior sense of smell and speed with the deadly cunning of human weapons in the hopes these two-leggers would prove trustworthy and share a piece of the action.

Many species faced with environmental challenges create useful, if unlikely, partnerships. There are so many examples of parasitic and symbiotic interspecies codependence in nature that it is possible our ancestors and the ancestors of our dogs may have been playing by the book. Badgers and coyotes, for instance, both eat ground squirrels. In challenging terrain they will team up to make their hunts more successful. This is especially interesting considering that the badger is no fan of the coyote, which has the nasty habit of eating baby badgers. Still, when a coyote comes calling, signaling his nonthreatening intentions through a series of mock attacks, jumps, and bows, the badger puts aside his distrust and more often than not accepts the invitation.

Ordinarily, lone coyotes rely on speed to snag squirrels that venture too far from their underground tunnels. A badger hunting solo can plug up tunnel openings and use its powerful front legs to dig out the trapped squirrels. But as a team, they combine their strength to greater advantage.

The badger will pursue the squirrel into its tunnel, and the coyote can then ambush it as it's flushed out. Sometimes the squirrel, sensing the waiting coyote, refuses to leave the hole and then the badger gets the prize. The success of their partnership seems to override the badger's hostility toward the killer of its young. It accepts its losses and collaborates with the enemy for the greater good. (It probably helps that coyotes seem to have no taste for adult badger meat.)

The badger-coyote relationship may start out as strictly business, but it becomes something else. When they're not hunting squirrels, these two unlikely allies actually rest next to each other and touch noses. This increased sociability seems to grow out of time spent on the hunt. The same thing appears to have happened when humans found out they and wolves had more in common than a taste for meat. Familiarity can also breed camaraderie.

Humans didn't just hunt like wolves; they lived like them as well. Humans lived in "packs" and cooperatively cared for their young. Besides being socially compatible, these two-legged carnivores offered another irresistible magnet. Humans cooked their kill, creating an olfactory sensation that some wolves probably couldn't resist. Those wolves that dared to approach discovered that these cooking animals threw away some of the best parts. (Early human hunters seem to have been more interested in marrow than meat.) Scavenging the rejects from early human camps may be what lured wolves into cave dwellings as long ago as 400,000 years ago. That's when wolf bones start appearing in the caves that humans lived in.

If some wolves entered human dens voluntarily, they must have been the boldest, most genetically predisposed of their pack. It would have been the least nervous wolf, or the hungriest, who made the most successful cave raids. If their lurking presence near camp kept more dangerous predators away, humans may have repaid this service by throwing these adventurous animals a bone. Soon they would decide not to stray.

Eating leftovers does not make a wolf into a dog, but it's a start. A bone or two would have sent a potent trust signal to the wolf, stirring her (pregnant?) social imagination. What we do know is that perhaps as early as forty thousand years ago, something environmental or social or both started some Eurasian wolves on their genetic journey toward dogdom.

The humans they were getting close to were also changing. They were developing the conscious awareness that would eventually make them the most powerful animals on earth. But for now, humans maintained their primitive ways and their "dogs" still walked and howled like wolves. On the surface everything seemed the same, except that both species found themselves more willing to get along. What started as an alliance of convenience grew to one of mutual respect and admiration—even affection. The strong interest we developed just watching wolves blossomed into a surprising sense of attachment when they finally let us touch them. When humans began to toy with raising young wolves, species distinctions began to blur.

Our female ancestors must have discovered that the cries of wolf pups elicited the same nurturing feelings that they felt for their own young. They instinctively reached down to pick up whimpering wolf pups, and these babies too quieted in their arms. It is not unreasonable to suppose that these women even breast-fed their adopted pups, according to animal behavior specialist James Serpell. In *Animals and People Sharing the World*, he explains that women in primitive societies all over the world have nursed infant animals. He quotes anthropologist W. E. Roth, who in 1934 reported with amazement that the tribeswomen of the Amazon basin "often suckle young mammals just as they would their own children; e.g. dog, monkey, opossum-rat, labba, acouri, deer." In eastern Columbia the Barasana women still suckle puppies and masticate plant foods to feed to their pet birds.

The emotional investment made in these animals has no connection to their economic usefulness; in fact, the greater the attention given these pets the less likely it is for them to become dinner or even barter. Even in societies that eat dogs, pups that are breast-fed by women are often spared because of the affection that wells up during the act of nursing. Roth also noted that breast-feeding had a profound bonding effect on the animals suckled as well: "Few indeed are the vertebrate animals which the Indians have not succeeded in taming." The story is much the same throughout indigenous tribes in South and North America, and there is no reason to think it has differed much throughout time. The infant cries, the nurturing touch, and breast milk all conspire to unleash the neurobiology of bonding.

Can such hormonal effects actually reach across the species barrier? Yes. Kerstin Uvnas-Moberg's team demonstrated that when humans stroke a rat forty-five times per minute for two minutes, they can raise the rat's oxytocin level. This gentle touch also causes a drop in the rat's stress hormone levels, heart rate, and blood pressure. Its behavior changes as well. The rat has a more sedated demeanor and is slower to pull its tail away from a hot plate, indicating that its pain threshold has also increased. All these effects can also be elicited by a shot of oxytocin.

When rats are stroked for five minutes, the effects are more pronounced and long-lasting. These extended touch effects mirror the rats' response to repeated injections of oxytocin. An injection of an oxytocin-blocking drug also stopped human touch from producing the analgesic effect.

The rats' antistress response is even more impressive when you consider that the control rats, who were held in exactly the same way but not stroked, had a significant and lingering increase in their heart rate and blood pressure. Not only did the massaged rats find human touch unstressful, but the stroking actually relaxed them.

It is also significant that these oxytocin-touch effects occurred in male rats. As we saw earlier, both male and female mammals make oxytocin, but females seem to have an oxytocin edge since estrogen increases oxytocin's release as well as the distribution and sensitivity of its receptors. The fact that the male rat's oxytocin system responded to gentle, rhythmic human touch shows just how responsive and inclusive oxytocin's calming powers really are.

As we will see in the next chapter, human touch can release oxytocin in another mammal because we all have nerves just under the skin that activate the oxytocin cells in our brains. So it is not surprising that our touch can produce such clear oxytocin effects in another species. Interestingly, Uvnas-Moberg found that the best antistress effects were produced at forty strokes a minute, the same rate we naturally use to stroke our pets. Great for the rats, but does petting also raise oxytocin in the petter? Although all the evidence pointed to this being the case, contact with pets has only recently been proven to raise oxytocin in humans.

In 2003, two South African researchers, Johannes Odendaal and R. A. Meintjes, published a study which showed that when eighteen men and women interacted with their dogs (talking to them and gently stroking them) the owners' blood levels of oxytocin almost *doubled*—and their dogs were also twice as enriched with oxytocin! Odendaal and Meintjes also showed that human-canine contact caused a significant decrease in the blood pressure of both, and the owners had a significant drop in their levels of the stress hormone cort as well. The dogs also triggered an increase in beta endorphins and dopamine production in their owners. Both of these neurotransmitters are also essential to our sense of well-being, and both are known to increase oxytocin levels.

These dramatic results could explain not just why we love our pets like children, but also what biological forces are producing the wide range of mental and physical therapeutic effects reported in animal therapy studies. These studies show that caring for pets can lower our heart rate, blood pressure, and production of stress hormones. For almost twenty years, we've known that oxytocin can reduce and regulate all of these physiological factors. So it is particularly satisfying finally to be able to say, yes, animals are releasing oxytocin in humans, and that's why they make us feel better.

Now, let's imagine the impact suckling a wolf cub would have had on our unsuspecting female ancestors. Permitting an animal to stimulate the ultimate oxytocin trigger—the mammary nerves—would have unleashed this powerful attachment biology in both the cave women and the wolf pups. Through innocent acts of kindness, Paleolithic women may have chemically cemented the most intimate interspecies bond we've ever known.

Clearly, great neurohormonal experiments were being carried out in the Ice Age. The shift in chemistry made humans more affectionate toward animals and created a more attentive and submissive wolf: in other words, a keeper. By the time we abandoned our old hunting grounds, neither of us was willing to be separated. The gentlest wolves—those most hungry for human contact—followed us into our new homes. The final transformation wouldn't have taken long once we began to selectively breed our favorite—most cooperative—wolves and help raise their young. How long would it have taken to turn the wolf into dog once we had become this intimate? If the wolves were anything like the Siberian silver fox, they may have been fully domesticated soon after the tamest ones began to mate.

The silver fox is a staple of the Russian fur industry. For over one hundred years it's been selectively bred for fertility and fur quality. These foxes are prolific and beautiful but not fully domesticated. Generations of captive breeding have had some effect on their DNA, but they are still fearful and aggressive toward humans. This is not surprising since they descend from an animal that has been described as a "bundle of jangled nerves" just as likely to run, attack, or freeze when threatened. This is a classic behavior profile of an animal with an overabundance of the adrenal stress hormone cort and low levels of the calming, socializing hormone oxytocin.

In the late 1950s Soviet biologist Dmitry Belyaev decided to try to make a better fox by breeding for a docile temperament rather than exquisite coat. To do this, he chose only the tamest animals, selecting pups that would not bite the hand that fed them or tried to pet them. The personality of these animals was closely monitored, but human contact was kept to a minimum in order to ensure that the domestication process was more nature than nurture.

After just five generations, the foxes being born were noticeably different. Some allowed experimenters to hand-feed them but showed no friendliness. Some were more social, wagging their tails and whining. By the sixth generation the team began to see the emergence of a "domesticated elite." These were the animals that Lyudmila Trut—the scientist who took over this study after Belyaev's death—says "are eager to establish human contact, whimpering to attract attention and sniffing and licking experimenters like dogs." They looked different too. These foxes had white markings, their tails curled, and their skulls were smaller. Their ears never achieved the smart pointed stance of the fox's, but remained droopy and puppy-like into adulthood. They even barked like dogs.

It's been forty years and forty-five thousand foxes since Belyaev first started selecting for tameness in his captive population. Out of all those animals, in such a short span of time and with so little human nurturing, has come the elite one hundred—each a product of thirty-five generations of selection for tameness. Trut describes them as "docile, eager to please, and unmistakably domesticated." Even when these animals escape their confines, she says, they all eventually return.

Belyaev's experiments revealed the same lucky secret that the first humans who bred wolves stumbled upon accidentally: nice wolves can have nicer puppies, and those nice puppies can have puppies that not only are nicer but look different. Even the blood of Belyaev's friendly foxes looks different. Their base cort levels have dropped 75 percent, and they produce far less when stressed than regular captive foxes. And their brains produce more of the feel-good chemical serotonin. The increased sociability as well as these physiological changes all indicate a bolstered oxytocin system. Oxytocin, however, was not one of the chemicals measured.

Belyaev discovered that tameness is a kind of genetic keystone that can support the entire physical and behavioral phenomenon we call domestication. But his work accomplished far more than making some foxes more doglike. The researchers who carried on Belyaev's work after his death developed close bonds to the tamer foxes. Their sense of attachment and responsibility toward their subjects grew so strong that they found themselves compelled to try to find good homes for the changlings when a lack of funding threatened to close the laboratory. Sadly, Belyaev did not live to see the ultimate result of his experiment: that making more affectionate animals makes more affectionate people. This is the biology of bonding in a nutshell. In a very short period of time, a single neurobiologically inspired human preference for affiliation could have changed the shape and fate of wolf, dog, and humanity.

Frederick Zeuner says that even the first generation of captive bred wolves shows a general reduction in body size. Another definitive skeletal change that tells an archeologist if he's looking at the bones of a wolf or a dog is the shape and length of the skull. Dogs have smaller skulls with shorter noses than their wolf ancestors. This may be another physical transformation that happened quickly. John Fondon and Harold Garner of the University of Texas Southwestern Medical Center in Dallas found that dogs have two sections of repetitive DNA sequences that control a gene called RUNX2. It is associated with craniofacial structure, and depending on the length of these repeat alleles, the dog will have a nose like a collie or a pug. This is the same sort of highly mutable genetic "tuning knob" that was able to change a rodent's social temperament in just one generation.

Sometime between forty thousand and fifteen thousand years ago, genetic tuning knobs started turning, and like Belyaev's foxes, wolves became affectionate to people and youthful in their disposition and physiology. They too came to lick the hand that fed them. They also retained their best wolf manners, offering their services to their new human family. As the most socially desirable wolves moved deeper and deeper into human society, they brought with them the wolf culture of respect and loyalty. Wolves need each other to survive. Belonging to a family is more important than their rank in the family. That wolf logic explains their devotion and submission to the humans they came to see as their parents, their betters. The primitive humans who became animal behavior experts would no doubt have noticed that their pet wolves treated them with the deference shown a pack leader. Among these humans, some would have figured out how to exploit this case of mistaken identity. For the first time, certain humans found themselves in a position of personal authority over an animal. They had become a kind of "domesticating elite," and it must have felt pretty good.

According to zoologist Dale Lott, it still does. Lott surveyed visitors to Rocky Mountain National Park where hand-feeding wild mountain sheep is a favorite pastime.

Lott was interested in discovering why so many tourists were drawn to engage in this close interaction with wildlife. Overwhelmingly, those who fed the animals told Lott that they wanted to see the animals up close and find out if the animals could trust them. When questioned further, they reported that they actually felt better about themselves when a wild animal would eat from their hand. They also said they thought more highly of other people who were trusted by these animals and that they felt their self-image was elevated in the eyes of others when a wild animal showed that it trusted them. Lott concluded that "for these people, feeding wild animals offers a significant opportunity to enhance their self-esteem."

It's fascinating to imagine our early ancestors' amazement and thrill at being able to interact in friendly ways with wild animals. After millions of years of being terrified by animals and then another million, or so, years of having animals running from them, these cave dwellers found themselves in a radical new social arrangement with animals that invoked a host of pleasing emotions. It's also amazing to realize that tens of thousands of years have not obliterated the pride and satisfaction we humans still feel when wild creatures see us as "friend." Whether a wolf or a human fired that first strange, flickering trust signal will always remain a mystery. But the robust "neuroeconomy" that grew up between the species is still with us today, and it is still why the trust of an animal can still signal our own trustworthiness to others.

Shakespeare knew that sometimes it takes an animal to tell you what lurks in the hearts and minds of men. As his King Lear stumbles across the stormy heath, he imagines he sees his dogs. But something's wrong: "The little dogges, and all, Trey, Blanch, and Sweet-heart: see they barke at me."

The dogs may be illusions, but they are still excellent judges of royal character. They deliver an unmistakable diagnosis—the king is not himself; his mind is not to be trusted.

The first humans to discover ways to make animals respect and trust them also inadvertently stumbled on a new way to befriend other humans. Not only did they manage to make wolves think better of them, but they ended up impressing their neighbors as well. This primal ego boost was so powerful that it inspired our ancestors to think "outside the cave." It wouldn't be long before we wised up to a whole new way with animals and with humans.

SEVEN

On the Shoulders of Giants

Our history with animals has been an ever-tightening spiral that's drawn us in from the remote wariness of watchers, to predators pushing the edge of the flight zone, and finally into an interspecies vortex where the distinctions of human and animal, parent and child, became forever blurred. There is no fossil record of the emotions that guided us on this intimate journey, but there is good reason to suppose that it was our oxytocin-inspired biophilia tendencies that ushered us ever closer to animals until they became ours—and we became theirs.

Today, it is impossible for us to grasp the role our long history of codependence on animals has played in the creation of our minds and our civilization. Just two hundred years ago, our genius with animals was still driving our intellectual, spiritual, and technological advances. Archeologist Stephen Mithen wonders if our hard-earned ancient mental talent for understanding animals and nature has been supplanted by our newfound fascination with technology. Is it possible, in this blink of time, to have so dramatically changed our hearts and our minds?

Not likely. E. O. Wilson insists that what he calls the "learning rules" forged in the natural world could not be erased so quickly—even for those of us who are several generations removed from the land. He has good reason to be optimistic, since there is ample evidence that those learning rules are not lost at all but safely stowed away, deep in the recesses of our primitive brain.

Evolutionary biology teaches us that nature adds far more than she subtracts. She's loath to jettison a good idea—and being able to bond with animals was a *very* good idea. So each of us still has a legacy of learning rules that whisper to us in childhood, dawn on us as we awake, occur to us in meditation, and are retaught to us from time to time by those who still have a way with animals.

For the longest time our interest in the horse was strictly epicurean. Judging from the number of horse joints found in prehistoric butcher sites, it appears we were hell-bent on eating them off the face of the earth.

Fair weather was no friend to the prehistoric horse either. Archaeozoologist Juliet Clutton-Brock says the horse was an animal strictly designed for the sweeping tundra of the Ice Age. As the last great ice sheets retreated, their pastures became overgrown with trees. The horse disappeared in North America, while the great herds of Europe and Asia dwindled in the face of starvation and predation by humans with their increasingly clever weapons and cooperative hunting strategies. By 9000 B.C., horses were well on their way to extinction when some of those hunters began to see this animal as something more than supper.

It would still be another four thousand years before some tireless watcher, probably from the Eurasian steppes, noticed something in some horse's behavior that made him suspect these were creatures that could be ridden as well as eaten. One of the first signs that there had been a shift in the human perception of the horse can be found in the remains of five-thousand-year-old settlements in the vast and beautiful grasslands of northern Kazakhstan. One of those sites, Botai, was home to a people who lived for and by the horse. There is evidence they had tamed a breed of horse probably related to the now extinct Tarpan. These stocky, smallish horses (thirteen hands) were kept in corrals and served as almost their sole source of food and milk. The abundance of leather-working tools found here also indicates that these people may have been among the first to make bridles and other tack used in riding horses.

It's hard to imagine the nerve it took for the first rider to test the radical theory of equitation or for some horse to allow it, but it was a dare that changed the world. By accepting their most dreaded predators onto their backs—their only blind spot—horses saved themselves and enabled those who could ride them to conquer the world—an evolutionary win-win by any standard. Simply being able to ride a horse allowed Alexander the Great to conquer 2 million square miles of the ancient world and the Mongols to create the greatest empire in history; the vision of a handful of mounted conquistadors brought the ancient civilizations of the New World to their knees.

That daring and highly imaginative human-animal partnership proved to be one of the most successful collaborations in human history and redefined the two species that engaged in it. The drama of this metamorphosis is replayed every day in the life of every horse and rider, and the quality of their détente remains a matter of life and death.

It's not surprising that horses were the last of the big mammals we managed to domesticate. Even today our horses remain one car backfire away from reverting to their primitive, wild selves. The horse was, is, and always will be a prey animal, with a nervous system designed to bolt or buck. Every time a human approaches an untamed or spooked horse, prehistory repeats itself. Which is why those who do it best still are a marvel to behold in the twenty-first century.

In the past fifty years, the American West has produced a new school of equitation often referred to as natural horsemanship. Ray Hunt is a high priest in this movement and a legend to horse owners around the world. For fifty years he's traveled from Maine to Australia trying to help people rediscover the art of communicating with horses. Horses talk to Ray. Actually they talk to us all, but Ray knows how to listen. That's why Nicholas Evans chose Ray Hunt as a basis for "Tom," the hero of his novel *The Horse Whisperer*. Whether on a horse or walking toward one, Ray can read its body and its eyes, and know what it's thinking and feeling. His powers of observation are a gift from a time when the tireless scrutiny of animals meant survival.

What Ray has learned (or remembered) after all these years is that horses have an overwhelming need to belong. A horse is a desperately social creature willing to sacrifice its freedom to be part of a herd. Ray uses this powerful bonding instinct to create a herd of two—himself and the horse.

Horses for all their might do not rely on brute force to create or maintain their social alliances. Mutual grooming helps keep the herd pacified and cohesive. When the social order does break down, they deal with it mainly through ritualistic threats and bluffs rather than violent confrontations. It is the horse's preference for subtle persuasion and nurturing contact that Ray Hunt relies on to establish what he calls a partnership between humans and horses. Ray knows firsthand that violence and aggression can successfully "break" these animals. And broken horses can be ridden. For many decades that's how he says he "got the job done." But he has come to see that those methods, based on anxiety rather than trust, at best produce only a limited partnership. Ray now offers the horse a better deal.

The sensibility that Ray Hunt tries to teach his students calls on ancient predatory skills. He needs them to rediscover their capacity for patience and acute observation—two mental talents trampled by modern lifestyles. This shift in perception is where it all must begin, or as Ray puts it, "It's what you do before you do what you want to do." It doesn't sound crazy to a horse.

Ray knows his watchful eye is no match for a horse's. This is an animal even more visually acute than we are. Ray's aware that once he's entered the round pen with a horse, his slightest move—even thought—will be noted and understood. This is his chance to "whisper," and it begins long before he ever approaches the horse.

He begins by thinking like a horse. This means acknowledging the fear demons that rule them. It also means entering their timescape. "Horses don't wear watches," Ray reminds his students. Ray's way with horses depends on his patient, generous thoughts and behavior. His goal is to have his body and mind work together to tell the horse the all-important message that it is Safe. He is not trying to break horses, he explains; he's trying to relax them or, as he says, to "turn them loose."

Ray has demonstrated his gentle method thousands of times all over the world. Each time it begins the same way: Ray and a "green" colt meet in a large round pen. At his approach the young horse bolts, running full-throttle circles around him. Ray knows this is part of the process and so he encourages it, waving the young horse on with an orange plastic flag tied to the end of a six-foot rod. He lets the colt experience its fear and react to it naturally through flight. Around and around the young horse gallops with Ray flagging it in different directions. Ray's in no hurry. He knows that thousands of years of domestication haven't erased running as the most basic learning rule of horsedom.

Soon the horse finds itself feeling more tired than scared. It also seems to dawn on the animal that this running business isn't getting him anywhere. He turns his eye toward Ray. "That's a good sign," Ray immediately notes. This fleeting glance—this hint of a "trust signal"—which is undetectable to most, is perfectly obvious to Ray. It may be the glimmer some sharp-eyed prehistoric hunter picked up suggesting that horses might have more to offer than a meal or two. What Ray does next may also be a reenactment of the first friendly human-horse encounter.

Ray begins to build on that very slight change by simply waiting and allowing the horse to consider some new options. What if he stops running? This guy with the flag hasn't done anything awful; maybe he can stop just long enough to catch his breath. The colt dares to stop on an angle while watching Ray out of the corner of his eye. His breath is heavy and his body tense and wet, but stopping was *his* idea. Ray lets him mull over just how this stopping business feels. Ray's seen it thousands of times: "This colt thinks he's in trouble and he'd better run, but running's a lot of trouble too, and pretty soon he'll see I'm offering him a better deal."

But not yet. This sideways stance is still defensive and poised for escape, so Ray snaps his flag at the horse's rump and sends him packing. "This is not a contest. I don't want him to protect himself around me." The process is repeated about a dozen times; if the colt still thinks he should run, Ray encourages him to do just that. If he thinks he can stop, Ray lets that happen too. Eventually the colt weighs the evidence: he's run and he's stopped—either way the guy hasn't hurt him—and at this point, stopping feels better. When he stops this time, he feels secure, even curious about this human. He turns his head and body to get a good look at Ray Hunt. Actually, it would be more accurate to say that the horse turns toward Ray to *give him a good look* rather than get one. The colt has been able to see Ray perfectly well all along. His big, beautiful eyes have almost total panoramic vision, a fact that must have created this early learning rule: Don't bother trying to sneak up on a horse. You only get near a horse if he lets you.

It's very much to Ray's credit that he's remembered how well horses see. There was a time not too long ago in Germany when a horse had to reteach us that primal lesson.

In the first days of the twentieth century, word began to spread about a horse named Hans who was able to do the most extraordinary things. His owner, retired schoolmaster Wilhelm von Osten, had taught Hans math, spelling, and music, among other things. A growing public was invited to watch von Osten put his horse through its academic paces in a tiny courtyard in Berlin. Facing his teacher and a large alphabet displayed in rows and columns, Hans dutifully spelled out the answers to a wide range of questions by tapping the number of the location of each letter. A correct answer earned him a snack.

The more who saw, the more who believed. The testimonials of witnesses spread the reputation of Clever Hans throughout the world. Respected experts from a variety of disciplines were assembled to verify the existence of the horse with a human intellect. The September Commission was composed of teachers, zoo directors, a circus manager, horsemen, veterinarians, and the noted psychologist Carl Stumpf.

The commission's 1904 report found no evidence of trickery or unconscious cueing and therefore had to conclude that Hans alone provided his answers. The European community was ecstatic, but some—including Stumpf—remained unconvinced and convened a panel to further investigate the cleverness of Hans. One panel member, psychologist Oskar Pfungst, was permitted to observe von Osten and others as they quizzed Hans on a range of subjects. He noted that the horse always wore a bridle and stood alone, untied or loosely held, usually by the questioner. The horse, according to Pfungst, "never looked at the persons or objects he was to count, or the words he was to read, yet he nevertheless gave the correct responses. But he would always make the most strenuous efforts to see the questioner."

Pfungst was also allowed to question Hans in private to eliminate any chance of coaching or signaling from von Osten or the small circle of people who were closest to the horse. Pfungst had noticed that Hans was most clever when quizzed by those "friends" so, as he later wrote, he also made a point to befriend Hans before testing him in isolation. Pfungst does not tell us just how he established his social bond with the horse, but he seems to have been doing something right. To Pfungst's amazement, Hans began to answer *his* questions correctly too. He was baffled but remained convinced that somehow the horse's correct answers had more to do with the questioner than the questions.

To test this theory, he had the horse quizzed by Hans's friends in the usual fashion. This meant that the familiar inquisitors posed questions that they themselves knew the answers to. Hans performed brilliantly, scoring 98 percent.

Next, Hans was presented with problems his questioners did not know the correct answers to. Hans's ignorance matched theirs, and he got only 8 percent of the answers right. Pfungst surmised that the horse's poor performance was due to the ignorance of the questioners and therefore, Hans's success also had to somehow reflect the questioner's knowledge of the correct answers. Still, he needed to figure out how information (or the lack of it) locked in human heads was getting transmitted to a horse.

Following his hunch that Hans was somehow "seeing" the answers, he had the questioners stand beyond the horse's line of sight. Again, the horse performed miserably. Now Pfungst knew where to look for the key to Hans's cleverness. After careful observations he concluded that Hans's questioners were involuntarily signaling the horse with the most minimal movements of their heads. This is how he described the process:

> As soon as the experimenter had given the problem to the horse he would involuntarily bend his head and trunk slightly forward and the horse would begin to tap. As soon as the required number of taps was given the questioner would make a slight jerk of the head, and the horse would cease tapping and return his foot to its original position.

Hans was relying on something called muscle reading. The horse could see the varying muscle tension in the tilt of the head of his testers. Pfungst further hypothesized that the success of Hans's response was not the result of mere passive expectation on the part of the questioner. The horse, as Pfungst had noted, responded most accurately to friendly questioners. Those who believed in Hans, Pfungst noted, "had a determination that he should do it." Pfungst believed that "an inward, 'Thou Shalt' as it were was spoken to the horse," most distinctly by those who felt an emotional connection to him.

Unlike von Osten, Ray Hunt is well aware that his body is telling his young horse exactly what he's thinking. He counts on it. Fostering a thoughtful and positive attitude toward the horse is the very thing Ray does before he does what he wants to do with any horse. Back in the ring, it's about to pay off.

The colt is coming to believe Ray's friendly body language. He lets Ray know this by turning his whole body toward him. That's the change Ray has been waiting for. He moves slowly toward the colt, his weathered hand extended with fingers bent. The colt is having second thoughts but manages to stay put as Ray calmly and gently rubs his face and neck. "He's finding out he can stand this, even like it."

Goodwill and good eyesight have gone a long way toward bringing horse and human together, but there are deeper magnetic forces at work as well. Ray's ability to walk up to horses and have them accept his touch is riding on the same biology of approach-interaction-relaxation that has negotiated all sorts of social interactions since mammals first appeared on this earth.

Sue Carter saw the chemistry at work in her prairie voles. She and her team created a kind of a mini round-pen experience for them in the form of a water-filled tub and a forced three-minute swim. The stress of the experience caused her male voles to seek the comfort and company of female voles. Female voles also found the experience a trial, but they preferred the company of other females to calm them down afterward. So, whether it's a rodent or a horse, stressful experiences, Carter says, "seem to increase social drive and interaction." When humans or other social animals are stressed, they too seek the comfort of friends or even strangers. When that comfort is received it produces the oxytocin that reduces the hormones that feed anxiety and replaces our dis-ease with a sense of well-being.

The round pen and Ray's presence in it create an interesting sort of stress test for the skittish horse. The animal is confined but not cornered. The human's flag waving is disturbing but not harmful. And then there's Ray's demeanor, which is quiet and calm. These mixed messages trigger a complex response that includes a rise in the fight-flight chemistry as well as an increase in oxytocin. It is the oxytocin that Ray must play to if he is to tip the balance of his colt's chemistry from defensive to accepting.

As we saw in Chapter 2, a rise in oxytocin urges animals to seek companionship. In the round pen, Ray is the horse's only social option, and oxytocin helps the colt recognize that it's a good one. Ray's deep understanding of horses and his genuine desire to bond with the animal informs his every thought and move, transforming the threatening atmosphere of the round pen into an environment Kerstin Uvnas-Moberg would call "nonnoxious."

Just as "noxious" events like pain or fear trigger the fight-flight reflex, nonnoxious, soothing contact, Uvnas-Moberg explains, causes a release of oxytocin and the complex chemical response of calm/connect described in Chapter 4. The result is a quieting of hearts and minds that allows horses and humans to "turn loose."

By sending powerful nonverbal messages of gentle confidence to the horse, Ray Hunt creates the trust and the oxytocin that defuse the horse's hair-trigger fight-flight defense system. This is no small trick when you consider that horses are prey animals, born to be hypersensitive to the slightest hint of threat. Their ancient history has "taught" them the bitter learning rule that we are out to kill them. This is the mighty evolutionary prejudice and chemistry Ray Hunt comes up against every time he walks into a round pen with a frightened horse. Fortunately for him and for the horse, he has stumbled onto something that turns off fight/flight: kindness.

Ray's nonnoxious thoughts and behavior send the visual signals that can unleash an oxytocin reaction capable of lowering the horse's vasopressin levels and the sense of territoriality and aggression that it can foster. With this defensive chemistry in check, Ray can begin to approach the animal. But it is his next act of touch that will seal the deal between him and the colt.

Oxytocin's bonding chemistry is orchestrated deep within the brain, but it often begins in the most superficial organ of the body—the skin—which is alive with countless sensory nerves. Each of these nerves comprises more than one thousand fibers of various sizes and textures that are individually designed to respond to different kinds of stimulation—especially tactile. So the bite of a whip or a spur sends pain signals racing to the brain via fibers known as C- or A-delta, where they trigger a defensive response.

Ray Hunt's gentle rub along the neck of his wary horse appeals to another class of nerve fibers—the A-beta type.

These nerves relay pleasant touch sensations to the brain where they release oxytocin and produce a sense of calm and connection. Not only can touch release a flush of oxytocin to relax a horse enough to permit human contact, the added dose of oxytocin that comes through gentle contact can stimulate opioid receptors and make the feel of saddle and bridle easier to bear. Ray's pleasant touch simultaneously releases the final oxytocin reward, which expands the initial chemical commitment of tolerance into a full-blown sense of well-being. It is at this stage that Ray Hunt declares a horse turned loose.

This is another bit of horse sensibility that our ancestors were well aware of. No, they wouldn't have known about oxytocin's role in the pleasure of touch, but the countless hours they spent staring at horses would surely have taught them that these animals (and all social mammals) like to groom each other. They no doubt figured out just how much horses rely on friendly touch to make herds and babies. Early hunters would have watched and waited while a herd indulged themselves in a bout of grooming, knowing that soon this swift prey would be lulled into a relaxed state that would make them easier targets.

World-renowned horse expert Claudia Feh has spent her life observing the habits of wild horses. In one fascinating study, she discovered that a horse's favorite and most elusive grooming site—its withers—is particularly sensitive to nonnoxious touch. It is here at the base of the neck, a spot best reached by others, that repeated stroking reduces a horse's heart rate—indicating an oxytocin-like inhibition of the sympathetic nerves. This is the magic power of touch that finally loosens up Ray's horses.

The kind of stroking that helps a horse calm down has been investigated in detail by Kerstin Uvnas-Moberg. As we saw in the last chapter, Uvnas-Moberg showed that gently stroking male rats raised their oxytocin and made them calmer and less bothered by noxious stimulation. Apparently this nurturing kind of touch was activating the A-beta nerve fibers that register nice touch and tell the brain to release oxytocin.

Uvnas-Moberg found that the calming response to gentle stroking was most powerful when the rats were petted on their abdomens. Using technology that can trace a single nerve fiber's route and activity, Uvnas-Moberg discovered she was also activating a new and unknown class of sensory nerves that seems to be uniquely dedicated to turning gentle touch into oxytocin. These special nerve fibers, found in the mammary abdomen and urogenital regions, bypass the standard sensory neural pathway that winds from chest and abdomen into the spinal cord and on up to the brain. Instead, these nerves travel directly from the torso to the root of the vagal nerve just outside the brain at the base of the skull. From here they follow the vagal pathway up into the brain stem to the hypothalamus where they can tap directly into the source of oxytocin. Touch-sensitive nerves with a special routing may have evolved to provide mother and baby with the most expeditious manner of releasing each other's oxytocin to ensure that they can create a symbiotic feedback system that will help them soothe and care for each other. But these same specific nervous pathways are also present and similarly effective in male mammals.

Uvnas-Moberg also found that petting (whether breast, vagina, belly, or back) is not the only type of comforting tactile contact that can tell a nerve to release oxytocin. Mild vibration and warmth also qualify as the kinds of nonnoxious stimuli that oxytocin cells respond to.

Once Ray Hunt has approached, touched, and relaxed a horse, it will walk right up to him. Even after thousands of years, it's still a pretty good trick. But taming a horse and riding it are two very different things, and knowing how to do one does not necessarily mean you know how to do the other. For one thing, all the reassuring visual signals you send a horse can only go so far. The irony is that once you mount a horse, you are sitting smack in the middle of its only blind spot.

As the first riders discovered, what happens next depends not on how you look to a horse, but how you *feel* to him. It is now up to your hands, your legs, your feet, and your seat to tell the horse it's okay and to ask for its blind faith. That tall order is just the beginning. Then you have to learn how to ride.

The human-horse riding experiment was first acted out in societies that were illiterate. We learn of this momentous achievement only thousands of years later when the descendants of those first equestrians rode into the classical world and astounded civilization. The Greek historian Herodtus tells us that Scythian nomads could not only stay astride galloping horses with nothing more than the simple aids of rope and blanket, they could aim their bow and arrow with deadly accuracy. The Greeks' shock and admiration were appropriate. Even the notion of equitation had not yet dawned on the West. They were witnessing not only the unthinkable, but the undoable. Were these men or some bizarre entertainment of the gods—half human and half horse?

These first accounts hint of equestrian powers invisible but mortal. The historical record tells us that these talented riders milked their horses, perhaps even suckled them. This nomadic practicality alone would have unleashed a powerful oxytocin influence over both man and beast, giving these horse keepers a leg up on riding.

Then there's the very act of riding—especially bareback—that offers the kind of tactile massage that can release oxytocin in both human and horse. For the rider, it would be the warmth and rhythmic stimulation of the genital nerves that would trigger oxytocin. And for the horse, the weight, warmth, and rocking movement of the rider massages nerves along the back and withers that can release oxytocin. The result was a coproduction of the chemistry that promotes trust, communication, and cooperation between social mammals.

The reward was mighty and mutual. Horses who could tolerate riders lived to eat rather than to be eaten, and the humans who could ride them instantly became bigger, stronger, and faster than those who couldn't. No wonder a profound bond was born between humans and the horse.

The skill of these first riders may never be matched again, but Ray Hunt is determined to try. In his clinics he strives to revive this ancient art one step at a time. He begins by asking his students if they've ever noticed how a horse will flinch when a fly lands on it back? Easy question, everyone nods. "Then why do you think we need spurs and whips and wires to communicate with such a sensitive animal?"

This is where things start getting tough. Next, Ray explains that when you're on a horse, its feet become your feet. "You must know at every second where your horse's feet are and what they're doing. When your horse's feet leave the ground, you can make those feet do anything you want. That's when you can ask them to speed up, slow down, cross over."

The riders take a deep breath as Ray goes on. "If you want your horse to walk a little faster, you just need to reach a little further with your legs, with the soles of your feet, the seat of your britches, and with your mind. You're picking up his feet and setting them down." Ray demonstrates by wrapping his reins around the saddle horn and folding his arms on his chest. Then, with no visible prompting, his horse moves out in a slow walk, picks up the pace, then drops back quietly. "See how little it takes to get the job done? It's the feel, timing, and balance. It can become natural as breathing."

Ray is trying to tell the world what it once knew so well. Historian B. K. Anderson acknowledged that such a connection between horse and rider was once the norm. "The ancient riders riding bareback or upon a saddlecloth, would enjoy a different means of communication with the horse, through seatsbones." Anderson laments that this degree of interspecies connection "nowadays is largely lost." Not to Ray Hunt.

This invisible communication between human and horse is as remarkable to witness today as it must have been in ancient times. Obviously Ray's supersensitive touch provides just the right pressure signal to fire the nonnoxious nerves that release oxytocin. With oxytocin levels heightened in both horse and rider, the chemical groundwork is laid that will encourage a level of mutual receptivity and cooperation that can only be called partnership.

There is another deep biological secret to Ray's way with horses, and it lies in his intimate understanding of the anatomy of their locomotion. When Ray Hunt's feet "become his horse's feet," he is actually acting out the horse's gait in his muscles and his nerves in ways that are invisible to us, undetected by him, but loud and clear to his horse. Ray, it turns out, is unconsciously propelling his horse by using the same subliminal physical forces that spell out messages on a Ouija board, spin tables in the air, and guide the hand in automatic writing. In fact, these psychic parlor games of the nineteenth century led to the discovery of deep neuromuscular powers that can turn all of us into master puppeteers.

The British physiologist William Carpenter was the first scientist to make the connection between thought and action. In 1852, he proposed that the mere idea or expectation of an action can produce, inadvertently, the motor behavior of that contemplated act. He called this invisible link the *ideomotor principle.*

For eighty years, water-witchers, Ouija-boarders, table-turners, and automatic-writers vehemently denied they were consciously or unconsciously manipulating their instruments. Finally, in the 1930s, physician Edmund Jacobson found how ideomotor actions get the job done. He placed electrodes on various parts of his subjects' body, then asked them to think about throwing a ball with their right arm. Amazingly, the muscles in the right arm registered the imagined toss. Similarly, merely imagining looking up at the top of the Eiffel Tower caused their neck muscles to contract. These experiments were the first proof of the marriage of thought and motion.

The fact that our muscles will execute an action—in little—from the slightest mental suggestion is just as incredible as the fact that this motion is impossible for us to control or detect. The psychologist William James summed up this eerie aspect of ideomotor action: "We are aware then of nothing between the conception and the execution. . . . We think the act, and it is done; and that is all that introspection tells us of the matter."

Jacobson may have exposed the deep and inseparable relationship between thought and action, but he did not discover it. Those who had already figured out how ideomotor action worked were making a tidy profit from it—the nineteenth-century mentalists who thrilled and amazed their audiences by reading their minds. After asking an audience member to select an item in the room and concentrate on it, the mentalist would take the person's hand, walk with him or her to the item, and identify it. Far from being guided by the subject's mind, the mentalist relied on the muscle tension in the subject's hand, which started out quite strong and diminished as the object came closer and closer. George Beard, the physician who first studied the extraordinary perceptive ability of these entertainers, more accurately described their art as muscle reading. And they couldn't hold a candle to Hans.

Psychologist Benjamin Libet was the first to bring the parallel universe of ideomotor action into sharper focus by tracing our actions to their neural roots deep in the brain before they play out in the body or the conscious mind. In the 1980s Libet showed that our conscious decisions to act are preceded by approximately half a second of nonconscious brain activity in the regions involved in coordinating motor activity, calling into doubt our identity as superior beings with free will. Libet suggested that this flurry of deep motor nerve activity might instead be offering what he called "free won't," a kind of trial run to see where our urges will take us and give us a chance to possibly override them. In 2008, Libet's study (which had measured neural activity via electrodes on the scalp) was repeated, this time using the latest in fMRI brain-imaging equipment. Under this more accurate and intense scrutiny, scientists were able to see that Libet's "free won't" actually begins seven seconds earlier.

Add to this the amazing discovery of the visuo-motor mirror neurons discussed in Chapter 1, and you get the picture that we are leading a very active prelife that is unknowable to us but visible through sophisticated technology or a trained eye. This is precisely the information that tells a predator where and when to attack, but it can also let an ally know when and how to approach.

Ray Hunt seems to be a natural-born muscle reader, and it's allowed him to enter and linger in the flight zone of another species. Besides being able to read horse muscle better than most, he can trick his muscles into sending messages back to horses as well. If merely the thought of an action can trigger its electrical execution, that faint muscular whisper can also be amplified by the depth of concentration and visualization applied to such thoughts. It is precisely this conscious manipulation of the ideomotor principle that Ray uses to send clear messages to the horses he's training. "I try to visualize my body and the horse's body as one.

When I'm first working with a young horse . . . I will sometimes exaggerate a walking motion when I'm trying to *let my idea become the horse's idea*."

Ray's concept of visualizing the movement as a means of transferring the idea to the horse is textbook perfect. In the 1950s, researcher Andre Weitzenhoffer discovered that "under favorable conditions, the minute responses initiated by thoughts can be built up to relatively huge proportions through neuromotor enhancement." This same mental talent that can make Ouija boards speak, divining rods point, and horses carry humans to glory will soon be animating artificial limbs. The "Luke arm," a prosthetic invention of Dean Kamen, has a range of motion and dexterity amazingly like the real thing, and the wearer will move it just the way Ray Hunt moves the feet of his horse. Kamer describes it like this:

> The patient thinks about moving the arm, and signals travel down nerves that were formally connected to the native arm but are now connected to the chest. The chest muscles then contract in response to the nerve signals. The contractions are sensed by electrodes on the chest, the electrodes send signals to the motors of the prosthetic arm—and the arm moves.

Ray Hunt is living proof that a single urban century has not paved over our genius with animals. He's reminding us, one ancient learning rule at a time, how to make our way with horses again. But there's something else that makes this thing called equitation happen, and it's something that can't be taught or learned. It's the biophilia of riding—the oxytocin factor—that makes it all possible.

Muscle reading and ideomotor action may articulate the intentions of humans and horses, but understanding each other's thoughts and emotions does not put anyone on the back of a horse. Simply put, we ride horses because they let us. Why they let us is complicated, but it has everything to do with oxytocin.

Riding is an act of faith that is based on a mistake. In fact, it is the ultimate case of mistaken identity. Anthropologist Elizabeth Lawrence has long been fascinated by the evolutionary coincidences that have conspired to make this flighty animal carry us into so many valleys of death.

> That a large, gentle herbivore, whose natural reaction to danger is swift flight, can be made to gallop forward into the noise and confusion of battlefields is testament to equine obedience to the human will. Herd instinct and adherence to equine social dominance orders may play a part in the cavalry charge. *But if this is the case, the rider in a sense takes the place of the horse's conspecifics, and conditioning and human mastery are superimposed on the animals.* [my italics]

Lawrence's insight about the merging of identities implies the involvement of oxytocin and vasopressin and their talent for social bonding and social recognition As we saw earlier, these neurotransmitters have the power to transform another gentle herbivore, the prairie vole, into brave and loyal mates and parents. In humans, oxytocin enables a mother to see herself in her newborn—a flash of recognition that will inspire her to protect it with her life. These two chemicals that flow through both humans and horses are capable of blurring species distinctions as well. Once that happens they can inspire a depth of mutual empathy strong enough to forge a Centaurian bargain, even in the face of death.

Today, riders no longer have to test their horses' commitment on the battlefield, but the essence of that faith still glows and permits every mount. It has survived centuries of brutal horsemanship to reemerge unscathed under more favorable conditions. Fortunately for both horse and human, we seem to be relearning the subtler and more effective ways of the first riders. We seem to be rediscovering the mental talents of conversing with horses. The conversations remain the same; the difference is that today science can shed some light into how they take place and why they are best conducted at a whisper.

EIGHT

The Meeting of the Minds

Once neurohormones nudged humans and animals together, the trick was to figure out a way to communicate. Fortunately our ancestors had the sharp eyes, the open hearts, and the quiet minds that were perfectly suited to the task.

We modern humans are the product of a brain and a vocal tract that can produce and comprehend sophisticated language. Words have become so dominant in our thought processes that it is hard to think of a life without them. But imagination and conversation do exist without words in a part of our brain that we once relied on far more than we do today. It's with this preverbal wiring that humans and animals came to know each other and concoct the intimate and essential social arrangements necessary for domestication.

All archeological evidence indicates that wolves were the first animals to enter human society. Wolf bones appear in hominid settlements 400,000 years ago. Those wolves and humans were hundreds of thousands of years away from becoming dog and master, but even then, the potential was there in the vague, mutual sense of recognition that flowed between them.

Besides relying on visual signals and expressions, early humans probably communicated by vocalizations similar to those used by wolves, so mutual understanding could have come about quite naturally. Wolf talk sounds a lot like singing to our ears. Human talk sounds a lot like singing to theirs. This is because we are both singing the same ancient song. And according to ornithologist Eugene Morton, it goes something like this: harsh, low frequency sounds, such as growls, signal hostility. Higher, pure-tone sounds, like cooing or whining, convey friendly and appeasing intention. This holds true for all mammals and birds as well. Morton calls it a "vocal convergence," and it's just the sort of ancient overlap that could have helped to cement the human-animal bond.

In the past 100,000 years, we humans have glorified our growls and whines into the powerful art and architecture of spoken and written language. This way with words is the consequence of a redesigned vocal tract and cerebral cortex. Precise neural connections in the left hemisphere of the human cortex created networks capable of inventing and comprehending words, and our new voice box breathed life into them. As far as we know, we are the only species to connect our neural dots quite this way.

Still, we retain a fluency in those expressive primal sounds and gestures that reveal emotions such as anger, fear, panic, sadness, surprise, interest, happiness, and disgust. Before we had language, we must have understood each other solely through touch, facial expression, the tone and volume of voice, the tempo of speech and breath, as well our body movements, both voluntary and involuntary. This rich repertoire of nonverbal expression is known as paralanguage.

The amazing thing about paralanguage is that it was not obliterated by the success of language. As Darwin noted, paralanguage remains our most reliable method of conveying our emotions and intentions to each other.

> Articulate language is, however, peculiar to man; but he uses in common with the lower animals inarticulate cries to express his meaning, aided by gestures and the movements of the muscles of the face. This especially holds good with the more simple and vivid feelings, which are but little connected

> with our higher intelligence. Our cries of pain, fear, surprise, anger, together with their appropriate actions, and the murmur of a mother to her beloved child are *more expressive than any words*. [my italics]

These sounds are still interpreted by our nonlinguistic brain with great accuracy and speed. In one study that examined the musical meaning underlying words, researchers taped conversations between patients and their surgeons—half of whom had been sued for malpractice. Then the researchers selected a mere forty seconds of these consultations and electronically filtered out the high-frequency sounds that distinguished the actual spoken words. What remained were the "vocalizations," the pitch, tone, and rhythm of the doctors' speech. Subjects were then asked to listen to these brief, altered tapes and rate the warmth, hostility, dominance, and anxiousness they heard in them. The surgeons whose tone of voice sounded most dominant to participants also happened to be the ones who had been sued. The doctors who were deemed less dominant turned out to be those who had never been sued. It seems that voice is an accurate and efficient instrument for sending a trust or distrust signal.

Nonverbal communication may have preceded verbal language, but it became inextricably folded into spoken language and critical to our comprehension of it. Nonverbal communication, philosopher Mary Midgely says, "supplies the context, and the only possible context, within which human talking makes sense."

Psychologist Paul Ekman says that a sharp eye can also shed light on the true meaning of any conversation—spoken or unspoken. Ekman spent seven years staring hard at the human face until he became an expert in its wildest behaviors. Ekman is less interested in the faces we put on and charts the expressions that appear unbidden and unknown to us. These are the looks that flash across our face and tell others what we're feeling whether we want them to know or not. There are, according to Ekman, about three thousand muscle movements in our face that answer only to our heart. He's catalogued these involuntary twitches and figured out which emotions give rise to them. He's also learned to spot them at a glance. Just as the Foré know their birds in flight, or Ray Hunt can spot when a horse turns loose, Ekman can see the anger masked by a smile from the slight tension it casts across the eyebrows. Most important, he's proven that all over the world—from Winnipeg to Borneo—people make the same faces and they mean the same thing.

This visual convergence of involuntary facial expressions extends to other animals besides humans, which is why our watchful ancestors were able to grasp the intentions and emotions of animals well enough to avoid them, hunt them, and eventually tame and care for them. The wild and dangerous world of early humans provided the motivation and opportunities that would help them become the greatest animal experts of all time. With their alert, preverbal minds, they learned animal secrets we may never know, despite our pets' best attempts to tell us.

Our pets are no longer wild. They are our captive audience and best students. Yet they have not lost their need to know us intimately. That knowledge has remained a matter of life and death to them. They have the prelinguistic know-how to get inside our heads and refrigerators. Our pets' rapt attention can be both flattering and disconcerting, leaving the distinct impression that they can see something in us that we don't know about.

What animals know about us, and how they know it, has been hotly debated since Clever Hans showed us what horses really "hear." Hans's owner, Wilhelm von Osten, thought he had taught the horse language, music, and numbers. Hans, it turns out, was not listening to his lessons nearly as closely as he was watching them. Hans noticed the muscle tensing and relaxing that accompanied von Osten's words and learned the muscle patterns that seemed to spell out "carrot." Clearly Hans was participating in a conversation with his human friends—it just wasn't the conversation they thought they were having. Interestingly, this "misunderstanding" consistently produced satisfactory results for both the humans and the horse.

Horses are large, muscular creatures with big, sharp eyes. They are born muscle readers, so it's not totally surprising that Hans managed to recast our words into a medium more user-friendly to him. But equine vision alone does not explain what happened in that courtyard in Germany. Hans performed best with those humans who had a sincere interest in his success—interest that may have expressed itself more clearly in bodily affirmations when the horse chose the "correct" answer. Also the questioners' earnest desire to communicate with Hans could have sent a reassuring trust signal that triggered the horse's oxytocin system and earned his undivided attention. A relaxed and focused Hans may have found it easier to read the unconscious messages broadcast by his questioners' muscles. In the end, Clever Hans told us not just what horses really hear, but what humans really say. He exposed the truth about human conversation: it still operates on a level as basic as the dialogue that passes between badger and coyote.

Animal lovers constantly marvel at their pet's ability to "read their minds" when it's really the owner's face and body that are being read with uncanny accuracy. This is the muscle-reading talent that we all must have relied on before words. It is also one of the mental talents that can be reawakened with tireless scrutiny or even a snort of oxytocin.

Researchers in Germany and Switzerland collaborated on a study to see if oxytocin could help humans "read minds." They gave thirty male volunteers a single dose of intranasal oxytocin before showing them pictures of faces on a computer screen that were cropped to show just the area around the eyes. The experimenters found that the men who sniffed oxytocin (as opposed to those who received a placebo) were significantly better at understanding the emotions being conveyed by the most subtle eye expressions. Considering how much the eyes tell us about a person's or animal's mind and heart, a boost in oxytocin during the Ice Age might have helped us see that first spark of social hope in the eyes of certain wolves.

It is well established that women are better than men at deciphering the social and emotional information embedded in nonverbal communication, possibly due to their estrogen-enhanced oxytocin systems. As we saw earlier, oxytocin helps mothers recognize their baby's face, smell, and cry. Oxytocin also increases a mother's ability to discern the emotional nuances contained in those preverbal utterances produced by infants—a sensory knack that permits them to monitor the well-being of their infant. Oxytocin also encourages the positive feelings that flow between baby and mother that are manifested and mimicked in an inviting range of facial expressions. These include the devoted gaze, the smile, the laugh, which are essential to the formation and resiliency of the mother-infant bond.

For all humans, conveying and accurately comprehending paralinguistic messages is essential to social well-being. An inability to read the meaning of a look, a stance, or tone of voice prevents autistic individuals from making rewarding social connections. Psychiatrist Eric Hollander of Mount Sinai School of Medicine in New York found that when he gave fifteen autistic patients intravenous doses of oxytocin, they showed clear improvement compared to the placebo group in their ability to comprehend moods such as anger, happiness, and sadness conveyed in speech. These effects were still apparent two weeks later when the subjects were tested again. This was the first study to show that oxytocin can help those with autism hear the emotion in spoken words.

Grunts, coos, or growls well up full of meaning that can be understood by all mammals. Psychologist Gregory Bateson described this communication overlap in his book *Steps to an Ecology of Mind.* Like Darwin, Morton, and Midgely, he found:

> We terrestrial mammals are familiar with paralinguistic communication; we use it ourselves in grunts and groans, laughter and sobbing. . . . Therefore, we do not find the paralinguistic sounds of other mammals totally opaque. We learn rather easily to recognize certain kinds of greetings, pathos, rage, persuasion, and territoriality.

These responses are so basic, so subcortical, that they are ingrained in all mammals and form the basis of our emotional understanding of one another.

Interspecies communication doesn't seem very far-fetched when you consider that human infants come into this world with the vocal equipment of a primate, and they grunt their way into language. An infant's tiny larynx attaches to its nasal passage so that it can only breathe through the nose. Over the next six months, the larynx descends and the throat opens, replaying the evolution of the vocal tract in each newborn. Even after the larynx has dropped deep into the throat, freeing the tongue to make a variety of sounds, grunts continue to make up a substantial portion of every infant's vocal repertoire.

Human grunts, like those used by other mammals, are the result of brief constriction of the larynx followed by the release of an abrupt vowel-like sound. At their most autonomic, these sounds are a by-product of the vocal tract closing during food ingestion to protect the airway. Grunts are created by the rush of breath released after it's been held during any strenuous act. Besides the effort of moving, reaching, and lifting, the very focusing of attention will produce "uh" grunts in infants. (Clever Hans watched for the "silent grunts" of his examiners.)

Like human infants, baby monkeys make their first grunts while struggling to roll over and find their mother's nipple. Primatologists Dorothy Cheney and Robert Seyfarth are famous for their interpretations of the vocalizations of vervet monkeys in Africa. They discovered that as vervets mature, so do their grunts. The effort grunts that accompany a monkey's first strenuous activities are later coopted for a new use: to indicate the movement of *others*, while the emitter remains quite stationary. This, child psychologist Lorraine McCune believes, is a direct transition from involuntary visceral vocalization to voluntary, referential communication.

After studying 4,161 vocalizations uttered by twenty preverbal babies between nine and sixteen months of age, McCune found a similar process at work in humans. The same vocal utterances created in response to metabolic demands reincarnate as effort grunts during strenuous movement or attention, and then advance to communicative sounds on their way to becoming human speech. This, McCune says, is how sound becomes meaning. She calls it the "sound-meaning correspondence."

Most recently two German primatologists discovered another vocal link between humans and primates: in tone and melody and volume we still express pain and pleasure much the way a squirrel monkey does. It seems that after 45 million years on our own evolutionary path, we have not lost our primate emotional "accent."

No matter how far we have traveled from our primate roots, grunts and grimaces expose our deepest thoughts and emotions. If this leaves us feeling somewhat vulnerable to the prying supersenses of our pets, we should take comfort in the knowledge that if we really try, we can figure out what they aren't saying.

Animal behaviorist Barbara Smuts found that by turning off her big brain, she was able to remember the ancient rules of social engagement that would allow her to enter into baboon society in East Africa. For twenty-five years, Barbara Smuts studied wild baboons. For two of those years, she traveled from dawn to dusk with "her" troop. She was just one of 135 primates moving across a vast range in search of food and safe shelter for the night. Mainly she lived alone, seeing only the baboons for days at a time. Even when she camped with other researchers, she was still living on baboon time: twelve hours every day, seven days a week. This total immersion into wild Africa, Smuts says, awakened ancient skills of intense concentration and fully aroused her senses as she entered into the zone of the hunter's trance.

Her relationship with the baboons was established one paralinguistic step at a time. She moved toward them in increments, stopping whenever they began to move off. As the baboons allowed her to inch closer, she was able to observe their subtle responses to her presence. Females signaled their young to stay close; more eyes followed her. As she began to read these social cues she learned when to stop her advance before the baboons became anxious enough to retreat. Her politeness was rewarded. Soon she was moving freely among the troop. This acceptance is classically referred to as "habituation." It is a term Smuts uses reluctantly:

> The word implies that the baboons adapted to me, that they changed, while I stayed essentially the same. But in reality, the reverse is closer to the truth. The baboons remained themselves, doing what they always did in the world they had always lived in. I on the other hand, in the process of gaining their trust, changed almost everything about me, including the way I walked and sat, the way I held my body, the way I used my eyes and voice. I was learning a whole new way of being in the world—the way of the baboon.

Besides learning to recognize more and more of the baboons' social signals, she was able to overcome her physical limitations enough to produce much of the body language necessary to engage in their life on their terms.

Smuts knew she had finally arrived in baboon society from the dirty looks she got as she moved closer to the troop. To Smuts these gestures signaled a victory because rather than being treated as an object to be avoided, Smuts says, she had become "recognized as a *subject* with whom they could communicate." Smuts had become someone to deal with, even educate, instead of someone to run from.

The first lesson was that each animal in the troop lives in the center of its own personal territory. The circumference and penetrability of this zone is a function of rank. Entrée to the higher ranked animals' private zone must be petitioned, usually with a grunt. That proto-word and the accompanying postures and facial expressions are loaded with information about the approacher's rank and intention.

Moving among a hundred baboons year after year is bound to land you in the personal space of many a baboon. Here Smuts found her university training to be a detriment. "Every field worker knows that it is critical not to move too close to the animals one is studying, so as to minimize one's influence on their emotions and behavior. . . . but ignoring the proximity of another baboon is rarely a neutral act." Smuts soon learned that brief eye contact or grunting—the appropriate social response in these circumstances—actually made her less invasive to the daily order of baboon life. Consequently she was able to make more detailed and accurate observations of her subjects' behavior.

Smuts's mastery of baboon communication skills and her attention to baboon protocol made her not only a better scientist but also a better baboon. "Over time they treated me more as a social being like themselves, subject to the demands and rewards of relationship . . . I was increasingly often welcomed into their midst, not as a barely-tolerated intruder but as a casual acquaintance or even, on occasion, a familiar friend."

Smuts spent most of her waking hours with the baboons following their routine, resting when they rested, sharing shade during the hot afternoons. Surrendering to the baboon way of life meant that Smuts had to suppress her own natural inclinations in various situations. For instance, when clouds were gathering, she invariably wanted to head for the shelter of the overhanging cliffs before the baboons did. And yet the baboons seem to better understand how to squeeze out the very last moment of grazing before running for cover.

Smuts says one day, out of the blue, this understanding came to her as well. She found that she just knew, like the baboons, when the first big drops would fall. "I had gone from thinking about the world analytically to experiencing the world directly and intuitively. It was then that something long slumbering awoke inside me, a yearning to be in the world as my ancestors had done, as all creatures were designed to do by aeons of evolution."

Smuts had discovered a dormant mental talent, a learning rule. Or more accurately, she was shown this way of understanding by a species for whom such awareness had never become redundant. The baboons, it turns out, had a great deal more to reveal to Smuts:

> The baboons' thorough acceptance of me, combined with my immersion in their daily lives, deeply affected my identity. The shift I experienced is well

> described by millennia of mystics but rarely acknowledged by scientists. Increasingly, my subjective consciousness seemed to merge with the group-mind of the baboons . . . I had never before felt a part of something larger, which is not surprising, since I have never so intensely coordinated my activities with others. With great satisfaction, I relinquished my separate self and slid into the ancient experience of belonging to a mobile community of fellow primates.

One of the most intimate and profound experiences Smuts recalls occurred during a blinding rainfall. Smuts had lost the troop in the downpour and ran for a old fishing hut on the shores of Lake Gombe.

> The inside of the hut was pitch dark, but I soon realized I was not alone. About thirty baboons were crowded into a space the size of an average American kitchen. When I entered, some baboons must have moved slightly to make room for me, just as they would for one of their own. But they didn't move far. Baboons surrounded me and some of them brushed against me as they shifted their positions. The rain continued. The hut filled with the clover-like smell of their breath, and our body heat transformed the hut into a sauna. I felt as if I'd been sitting this way, in the heart of a baboon circle, my whole life, and as if I could go on doing this forever. When the rain stopped, no one stirred for a little while. Maybe they felt the same contentment that I did.

And most likely they did. In fact, it would have been hard for the baboons to feel anything else. They, like their human friend, were under the influence of the same forces created in that rain-soaked hut. The warmth, the closeness, the clover scent, the rhythm of the rain, the camaraderie would have appealed just as strongly to their mammalian brain and filled them too with oxytocin and its gift of contentment.

It's not the only time oxytocin intervened in Smuts's relationship with these animals. Of course it was oxytocin that made Barbara Smuts want to live with baboons in Africa in the first place. It's what made her want to understand and be understood by them. And it was oxytocin that finally allowed her to see herself in these others. That is oxytocin's other great gift, the capacity to merge. What happened to Barbara Smuts happened because oxytocin helped her access her ancient brain and its animal-like brand of attention. Our ancestors never knew any other brain. Theirs was the genius of living in the here and now, and their here and now was filled with animals.

They were at their nonlinguistic best when they first experienced the kind of mind meld with wolves that Barbara Smuts felt with baboons. Call it beginners' luck, but they had stumbled on the one species that understood their growls and whines best. They had had met their paralinguistic match, and just watching animals would never be enough again.

NINE

The Dog of the Hare

Canines don't just speak our paralanguage; they share our carnivorous outlook on life. Our relationship with them has proven to be so successful and enduring because it was conceived in common terms and in our common interests. Canines and humans looked at those great herds and saw the same thing: dinner. We also divided loyalties and duties in similar ways. Our notions of kinship and protection were as compatible as our menus. With dogs we shared the common language of survival.

Today, dogs needn't help us run down dinner, but many still can and would if they were asked. They have, however, remained most attentive and responsive to our changing needs and wishes. And what humans need now, more than a good hunting partner, is someone to talk to, someone who will listen to them.

Dogs have excellent conversational skills. Like horses, they are brilliant muscle readers. Even if they never learned a single command, they would still know when you even *think* about getting up, putting on your shoes, or taking a W-A-L-K. Your ideomotor muscle action spells it out for them. And they have a unique ability to understand our deliberate gestures as well. Dogs—not wolves and not even chimpanzees—naturally follow our gaze, our pointed fingers, even the tilt of our heads. Animal behaviorists wondered if the dog's ability to follow a glance or a gesture and comprehend its meaning came about because our ancestors selectively bred wolves with the talent for it.

Two primatologists, Richard Wrangham of Harvard and Brian Hare of the Max Plank Institute in Leipzig, Germany, teamed up with Lyudmila Trut to see if her fox-dogs had developed this insightful capacity. Trut's tame fox population had never been tested for their ability to read human communicative gestures, but they turned out to be just as good as dogs at reading our paralanguage. A control group of foxes not bred for tameness failed to pick up on these paralinguistic cues. Since Trut's domesticated foxes were not bred for their sociocognitive skills—only for their tameness—the team concluded that their capacity to attend to the communicative attention of humans arose as yet another "by-product of selection on systems mediating fear and aggression." Brian Hare believes that wolves actually have these cognitive capacities as well, but their fear and aggression mask them. Oxytocin has the power to strip off that mask and create a temperament that allows dogs (and fox-like dogs) to be calm and caring enough to follow our intentions very closely.

Dogs can follow our words as well. Your dog may not be able to "talk," but he does know the meaning of the W word as well as car, out, come, stop, no, okay, his name, and at least sixty more words. At a minimum, dogs know the music of these words and many more. You wouldn't want to compete with this animal in a game of Name That Tune.

And your dog learned these words just the way you and I did, by building on the lullaby of the classics: come, stay, good, bad. These four words are the cornerstones of social language for all mammals, and they are chained to a universal melody that we are all born humming. Parents all over the world use the notes of this song to charm and instruct their babies. This is the very basic human language we call "motherese" or "parentese." It also happens to be the same acoustical formula that animal trainers worldwide find so effective in communicating with their mammalian students.

Animals, like humans, recognize the intention of a voice command from its acoustical profile. Prohibition, for instance, comes out in the evenly staccato notes and pitch of "No, No!" "Lassie come home!" is best heard in a series of four ascending notes of encouragement. Comfort is delivered on the descending pitch of "There, there." Attention and approval are given with notes that quickly rise, hold, then fall sounding a lot like "Good dog." Using this simple scaffold of paralinguistic understanding, humans and animals began a wordless dialogue that inspired trust and loyalty in both.

As our wolf-dogs became tamer and more emotionally connected to their human pack, their ability to read our muscles, our behavior, and our rudimentary vocabulary must have improved. The more comfortable they became being near us, the easier it would have been for their finely tuned ears to also pick up the words we were adding to our instructive melodies and gestures. This may explain why, after tens of thousands of years of living and evolving with humans, a dog named Rico now understands the meaning of two hundred words—and then some.

Rico the border collie was born in Germany in 1994. When he was around ten months old, his owners began playing a game with him that involved hiding stuffed toys throughout the house and asking him to retrieve them by name. He was quite good at this, so his owners kept buying toys, naming and hiding them. By age ten, Rico knew the names of two hundred different toys and could fetch them on command.

Juliane Kaminski, Josep Call, and Julia Fischer of the Max Planck Institute for Evolutionary Anthropology in Leipzig, Germany, first learned of Rico when he was featured on a television show. Rico's owners agreed to bring him to their lab to have his language skills tested under controlled conditions. In the first study, the experimenters tested Rico's comprehension of the two hundred items the owners said he knew by name. They placed twenty sets of ten toys in a room. Then they had the owner ask Rico to fetch items they had randomly selected. Rico went alone to the adjacent room and returned thirty-seven out of forty times with the correct item. This put Rico in the same language comprehension league with language-trained apes, dolphins, sea lions, and parrots.

The experimenters then added a twist. They placed a brand-new toy with seven toys Rico knew by sight and name. The owners then asked Rico to get this totally unfamiliar item. Despite the strange name—or perhaps because of it—Rico was able to deduce that the unknown name being requested must belong to the strange item among all the known entities. Rico brought the right toy back 70 percent of the time, showing that he has a talent for something called "exclusion learning." Four weeks later, Rico returned to the lab and was asked to fetch one of the novel items he had retrieved in his last visit. This item, which he had only seen that one time, was now placed in a room with four familiar toys and four novel ones. He was correct 50 percent of the time, the same accuracy rate seen in three-year-old children. (When he was wrong, it was because he retrieved another one of the novel toys, never an old familiar one.)

Border collies are bred for their ability to comprehend human communication signals as well as for their work ethic. The scientists who studied Rico all agree that he is a highly motivated dog. His desire and ability to focus on his owners' every word seems to have revealed to him the learning rule: "things can have names." His intense motivation and sense of reward for matching item to word also quickly and firmly locked them into his memory. These are the same cognitive abilities we call "fast mapping" when our toddlers do it; without it they would never be able to acquire language. But the German team believes that Rico has shown that fast mapping is not necessarily a linguistic skill unique to humans. Instead, they think this may be a kind of associative learning that is practiced by other species. Consequently there is yet another shared mental talent that facilitates communication between us and the canines in our lives. Harvard psychologist Paul Bloom sums it up like this: "Dog owners often boast about the communicative and social abilities of their pets, and this study seems to vindicate them."

There's also some scientific vindication for all those dog owners who've always suspected it is they who are being schooled by their pets. Miho Nagasawa of Azabu University and his colleagues in Japan observed how fifty-five dog owners responded to their dog's soliciting stares. They found that those who returned the look most often and longest were the same owners who said they felt the strongest sense of communication with their dog and the ones who came away from the test with significant increases in oxytocin. This finding becomes particularly interesting in light of recent studies that showed that oxytocin makes people spend more time focusing on a person's eyes and that it helps us read the meanings conveyed through eye expressions. This means your dog can chemically inspire you to look deep into his big wet eyes, find the meaning in them, and throw the stick, even as you are saying, "no."

Dogs didn't just learn to be very good watchers and listeners. Dogs, like human babies and vervet monkeys, have done an excellent job of completing Lorraine McCune's sound-meaning correspondence with "the bark." Yes, much to the annoyance of apartment-dwelling Homo sapiens, dogs bark. Perhaps the bark would be a little less vexing if it could be heard in an evolutionary setting. First, wild adult wolves rarely bark except to alert the pack and establish territory. But wolf puppies bark readily. The dog's facility for barking is one of those aspects of arrested development that turned the wolf into a pet.

Linguist Mark Feinstein is puzzled by the bark. He sees no immediate biological payoff for the phenomenal amount of energy it takes to produce the sound that penetrates your wall. But if we turn back the clock to the days when lupines were learning how to be dogs, we can see how the bark might have become a very important part of the dog's dowry and its entrée into human dens. If a dog's bark could dissuade an attack by a predator or, at a minimum, warn dull-sensed humans of its presence, the bark would be a very welcome sound indeed.

Even today, dog owners say they feel safer and sleep better in the presence of their dog. It's easy to imagine that the great dangers of the Ice Age denied the first dog "owners" much in the way of sound sleep. The defensive chemistry that kept them alert to threats is not easily turned off. Just knowing their dog could and would bark may have allowed human dog-keepers to rest assured for the first time. The gift of a good night's sleep could have been enough to secure a spot by the fire for these new noisy companions, making barking one savvy survival strategy. Barking, in fact, may have been the best trick the dog ever came up with.

Barks may have made our ancestors not just safer but smarter. Sleep is essential to the mental and physical well-being of all mammals. It is, of course, a time to repair and restore cells and energy used up in our waking lives. It is also a time of learning, when lessons taught in the day are repeated until they are etched into the neural circuitry that will allow us to retrieve them as memories rather than have to relearn them each time. With a dog on watch, our ancestors could finally let their vasopressin guard down and slip into a deeper sleep that is accompanied by long, slow alpha-wave brain action. This state of decreased sympathetic nerve action and increased parasympathetic activity supports the complex learning of behaviors and language. This is just another reason to agree with the anthropologist Claude Lévi-Strauss, who said that "animals are good to think."

And last, but certainly not least, our dogs have made the most of this adolescent vocal behavior by expanding the guard bark into a greater range of meaning. They have adjusted the tempo and tune of their barks in a dozen ways to signal their need for food and friendship, to come and to go, using the same emotional melodies that we use to underscore the same sorts of messages. In a fascinating audio experiment conducted in Hungary, it was shown that humans—whether they had ever owned dogs or not—were able to correctly distinguish the emotional state of a dog from listening to a recording of its bark. Far from being a loud waste of breath, barks comprise acoustic structures that make them understandable to human listeners.

Belyaev showed that with tameness comes the bark. Peter Prongracz and his colleagues in Budapest showed that barks are meaningful noise to our ears. Feinstein is even willing to grant them the status of protoword. Like the primate grunt, barks occupy that gray zone in which sound becomes concept, offering a hint at how human language may have appeared.

Another clever canine linguistic trick is the play-bow. With one simple gesture—front legs splayed, head down, butt high—dogs transform a host of sexual and aggressive body language and vocalizations into playful, benign social behavior. It's doggie double entendre—a variation on the coyote's harmless proposition to the badger. Once the play-bow establishes good intention, a more generous translation of these classic defense expressions is possible. Still, the noisy wrestle of locked mouths and front legs is just an instant away from reverting to its original defensive meaning. Just below the surface of dog play, oxytocin and vasopressin vie for the lead in this intimate tango. Vasopressin employs these gestures to signal territoriality and aggression, but with oxytocin out front, the tone of the acts is not perceived as sinister. This allows dogs to use "bad" language to engage in a wider range of social communication with other dogs and with us. And then there are the half dozen meaningful variations on the growl and the different ways to wag a tail (left bias signals friendship, right is caution). Throw in some whining and intense glances from owner to desired object (ball, door, food) and a paralinguistically astute dog owner has no trouble holding up his or her end of the conversation.

Before language, we developed the expertise to read a face or catch the meaning in a melody, regardless of who made it—human or animal. It was this mental talent, says, Juliet Clutton-Brock, that helped us to negotiate domestication:

> Non-verbal communication . . . is very highly developed in human beings. . . . More readily than from his speech we can tell the mood of a neighbor by an often unconscious assessment of his facial expression, posture, and bodily movements, and *in the same way we can interpret the behavior, attitudes, and feelings of many members of the animal world. It is this ability that has greatly helped man to enfold a wide variety of animal species within his own social organization.* [my italics]

Words may have been unavailable when we were striking up our first conversations with wolves. Fortunately, Mary Midgely says, they were also unnecessary because "making oneself understood is an immensely wider field than talking." It is in this wider field of protowords and paralanguage that we find we do indeed have a fair amount to talk about with animals and a surprising number of ways in which to carry on that conversation.

Humans most adept at paralinguistics found themselves with willing hunting partners. Like the coyote, these human hunters were able to stop chasing their prey and let the wolf-dogs deliver it to them. Watching their canine partners drive a herd of ungulates must have inspired the hunters' dream of herds of animals just theirs for the picking. The humans who best conversed with wolves eventually managed to instruct them to round up their prey but not kill them.

The capacity of some wolf hybrids to inhibit their killer instinct hints of oxytocin's rising influence. A sense of recognition emerged in these changing wolves that prevented them from seeing weaker, helpless creatures like humans or sheep as fair game. Now they were inclined to gather up whatever animals we pointed to just as they would their pups. They watched over our herds and kept them and us from harm. This shift from infanticide to parental behavior is the same conversion that oxytocin and vasopressin inspire in rodents. It's the kind of personality change that gets licked or loved into babies. Generations of human-reared—even breast-fed—wolf pups could have trotted away with brains loaded with oxytocin and vasopressin receptors in all the right places to help them merge with us.

It took hundreds of thousands of years for humans and wolves to understand that they could approach, interact, and finally relax in each other's company. The calmest, most conversant wolves would become dogs. As these most socially desirable dogs talked their way into our homes and hearts they were also securing their place in evolutionary history. They fostered the human friendship that would grow their species some 400 million strong. Their pack mates who shunned our company have seen their world shrink until fewer than 400,000 wolves remain.

The ability of canines and humans to make themselves understood to each other profoundly and irrevocably changed the canine world. Humans were transformed by those ancient conversations too. To get a sense of what happens to humans when dogs come barking, we can tune into a human-dog conversation that was struck up relatively recently just below the Arctic Circle.

Dogs probably crossed over the Bering Strait into North America with the first human explorers, but, like most of their two-legged companions, headed south for warmer, easier times. The humans and canines who remained in the north became a breed apart. They were bonded to each other by a climate that recognized only two kinds of species, the weak and the strong, and the belief that they descended from a shared spiritual ancestor—the wolf.

For the Hare clan of Canada's frozen Northwest Territories, this kinship with dogs was more mythical than practical. The frozen forests and lakes that surround the Hare comprise one of the meanest environments on the planet. Starvation is top predator in this world, and one that dogs could not protect them from. Only the most well-off families were able to afford a small dog team. For the rest, dogs were a liability. The heavy work of pushing and dragging sleds across thousands of miles of ice each year fell to a more economical beast: woman.

Like the hunters turned herders, the Hare became entangled with dogs due to a simple career change. In the mid-1800s, fur became the rave of European fashion. The industry turned to Inuit tribes to provide a steady stream of exotic, lush pelts. These commercial demands could not be met by the traditional subsistence hunting methods of the northern tribes. Hundreds of miles of traplines would have to be set and continually monitored if this frozen landscape was to satisfy the European quotas. While the first farmers had to settle down in order to coax a harvest from their fields, the Hare had to commit to a whole new kind of migratory lifestyle in order to increase the yield of their land. This meant traveling many more miles, and at a hustle they had never known before.

Tending their vast traplines required greater efficiency and speed than the traditional meanders from camp to camp that had previously defined their lives. For the first time, women were no longer the best way to get the job done. If the Hare were going to make it as trappers, they were going to have to reintroduce themselves to their ancient canine kin and try to remember the learning rules that would give them a way with dogs. The Hare may have had to play catch-up in the dog bonding department, but, like the Foré, they still had the mental talent suited to the task. They weren't big talkers; they were keen observers.

For the Hare, the Ice Age never ended. They've always lived out on the ice with just primitive tools and their superb senses to keep them alive. Anthropologist Joel S. Savishinsky remembers traveling with the Hare one forty-below morning. He marveled how, without the aid of thermometers, they could detect temperature changes of five degrees and were constantly aware of the strength and direction of the wind as well as the quality and depth of the snow:

> Like a computer digesting a dozen sources of raw data, the people could process the environment with simultaneity, precision, and a flexible view to multiple outcomes . . . one could not help but sense and respect the experience, balance, and judgment that went into every hunting and traveling decision. Yet it was all done by the Hare with an outward quietude—a few words, silence, a nod or two—and the men had reached a consensus on a good route, a reasonable timeframe, and a fruitful objective.

It's hard to imagine any people who could better appreciate the dog's superior sense of smell, sight, and hearing. Dogs have been called sensory extensions of the human nervous system. Certainly, the dendritic anatomy of the human-dogsled team takes that observation well beyond metaphor. A team provides a "peripheral nervous system" that communicates incoming sensory data down the leather harness to the driver who then assesses the input and signals back to the dogs accordingly. With four paws on the ground they can feel the depth of the ice, but it is their sense of smell (dogs have about 220 million scent receptors, while humans have fewer than 5 million) that tells of changing weather or the approach of an animal or human even in a total whiteout. In this treacherous landscape an animal that can stay the course and find the game more quickly and accurately than the driver would be worthy kin indeed.

It's also not surprising that such sensitive, quiet people would become experts in the psychology and behavior of their dogs. In no time the Hare were able to recognize which pups had the character traits necessary for a "lead dog." They even became adept at developing that animal's leadership skills and matching the talents and temperaments of the rest of the team to the personality of the dog they must learn to follow. Together, a good Hare driver and his lead dog are able to transform a pack of eight aggressive, unruly animals into a living unit with a single vision and purpose. Not only are good lead dogs a matter of life or death, they bring their owners the respect of the entire village. The cohesion, appearance, and stamina of a team speak volumes in this taciturn society. "Among the highest compliments that a man or a woman can earn is to be considered a good hunter and 'good with dogs,'" Savishinsky said.

The Hare's prosperity and pride quickly became harnessed to their team. As trappers, the Hare covered greater territory but in a more circumscribed pattern. Their cyclical, nomadic life became defined by the circumference of their traplines. As we saw in Belyaev's study, selecting for one particular trait can deliver a host of unimagined side effects. For the Hare, the decision to hunt with dogs ultimately changed everything in their lives. As Savishinsky noted, fur trapping with dogs "both increased general mobility while diminishing people's level of nomadism." Families who had always followed the hunt now stayed in camp and had meat delivered to them by dogsled, while the men began the endless task of hunting and fishing in order to feed the growing number of dogs. By 1976 the Hare were seriously outnumbered—three to one—by their dogs.

We can only hypothesize about the impact animals had on every aspect of our prehistoric lives, but with the Hare we can witness how fast and how easily humans can become both master and slave to an animal. Besides dictating the work habits of their masters, dogs quickly took over their leisure conversations as well. These stoic people are at no loss for words when it comes to talking about their dogs. A dog's color, weight, coat condition, personality, and character within the team became the Hare's major topic of conversation. When they do speak about a person, his lead dog is always mentioned.

Emotional displays are rarer than words in subsistence cultures like the Hare's. And yet the dogs of the Hare have been granted the affection once reserved for their children. Why? It's simple economics. Affection in the constant face of death is the most costly of investments. To love is to cling to life. But the Hare seem to instinctively recognize that with kids and dogs the rewards of love outweigh the risks. Despite their recent arrival among the Hare, dogs have grown essential to both their physical and spiritual well-being. Just how deep has the bond grown? In a landscape that can be so cruel as to force people to eat each other, dog is simply not eaten—ever.

Even with their fantastic understanding of the laws of nature, the Hare never could have predicted—or avoided—the depth of commitment they've come to feel toward their dogs. By the end of the Ice Age, the human-wolf association, while still one of convenience, must have been becoming emotionally satisfying as well. Humans must have come to feel a parental sort of pride and joy in their dog "offspring." As human and wolf came into more intimate physical contact, new feelings must have stirred, accompanied by dramatic physical, mental, and behavioral changes in both. The chemistry flowing between the species was so strong it turned wolf into dog and humans into herders and breeders. What likely started as simple act of maternal affection grew into the revolutionary concept of animal husbandry—a misnomer for sure.

Like the Hare, we had unknowingly hitched our fate to the dog. They became our biggest fans during our hardest times, and soon we couldn't bear to be without them.

Twelve thousand years ago a human hunter was laid to rest in a grave with his hand gently caressing a puppy. The tiny bones could be wolf or dog or something in between, but this little canine creature was already capable of providing comfort—in this life and the next.

TEN

New Game

Once we started keeping dogs, we began to develop a more nurturing personality and the best friend that would help us fold wild plants and animals into the human society that was about to be born.

It happened first in the Levant, a two-hundred-mile-wide stretch of land that borders the eastern Mediterranean coast from the Sinai peninsula, north through Israel, western Jordan, Lebanon, and Syria, ending up against the Taurus mountains in Turkey. It was here, around 14,500 B.C., that the world first blossomed in the great spring that signaled the end of the Ice Age. The Levant had always had its oases even during the worst of times. Its ice-free coastal wetlands and mountain watersheds supported oak woodlands and meadows teeming with animal life. But in 12,500 B.C., temperatures grew warmer, ice melt started flowing, rain started falling, and the surrounding desert scrublands sprouted lush short-grass prairies. Land that had always been uninhabitable now bore over one hundred kinds of edible fruits, seeds, and tubers. These blooming Edens attracted humans who brought the dogs they were finding it harder and harder to part with.

Together they stalked the wild boar, deer, goat, aurochs, and gazelle that flourished in the new, improved Levant. So plentiful was the bounty of wild game, shellfish, eggs, nuts, seeds, and grains that for the first time, humans wanted for nothing. So these migratory hunter-gatherers tried something radically different: they stayed together and stayed put.

Loose affiliations of about 150 people built small, partially submerged wood huts or shelters within caves. They also erected larger buildings where they could gather together to celebrate their good fortune in ritual feasts. Fine weather, flowing water, and fertile earth must have offered the first profound sense of relief experienced in human history. The world had become a kinder, gentler place, and the neurochemistry of the post–Ice Age people must have responded to it. A growing sense of well-being could have tweaked genetic tuning knobs in favor of more trusting, generous, and cooperative temperaments. In these relaxed, receptive times, the social imagination began to expand beyond ancestral or even species boundaries.

The social chemistry in these first settled communities was put to the test in 11,000 B.C., when a long, cold dry spell came like a sobering slap in the face. Plant life became less productive, and waving grasslands were reclaimed by tougher drought-resistant vegetation. Game became scare and some communities had to be abandoned. The more established settlements with access to good water fought back. People came together and cleared the encroaching scrub. They tilled the hardened land and replanted the grasses on which they now relied.

These first attempts at farming did not require a great leap of imagination, according to Colin Tudge. In *Neanderthals, Bandits, and Farmers*, he proposes that people had been sowing and harvesting wild grain for tens of thousands of years. In his hypothesis, even the most minimal horticultural efforts could have proved valuable. Scattering seed onto the muddy soil of a brief Ice Age summer, says Tudge, could have produced a filling meal when hunters came home empty-handed. Or maybe, he thinks, these Ice Age gardener-hunters realized that sowing seed would bring forth the sweet grasses that lured the herds to them.

Either way, Tudge says, this sort of rudimentary farming would have given humans the ability to recognize a good grass when they saw it and the wit to know what to do with it. He thinks that the sickle, a new technology that appeared in the settlements of the Levant, indicates that people were now cutting wild oat and barley grasses in great swaths to be brought back to the village and threshed. The violence of the mowing and transporting would have knocked off all but the largest, most stem-fast seeds. Such seeds naturally hit the ground last or not at all, but now they became the only ones that humans would hand deliver into prepared soils.

Before long, according to Tudge, the bulk of the grain growing in the Levant came from these tenacious seeds that matured into plants loaded with seeds that could be released only with a good threshing. In an incredible bit of good fortune, the larger seeds the early farmers "selected" produced a better class of grain.

By 7000 B.C. the glaciers had been wrung dry, and the great forests that surrounded the Mediterranean dried up. Farther north, trees, like giant weeds, began to cover the warming tundra. Countless herds of grazing animals hugged narrowing rivers or followed grasses at the edge of the retreating ice. The settled communities of the Levant found themselves devoting more energy and consideration to growing grains. The preparation of the soil and cultivation of new grasses became full-blown farming—a far greater cooperative and conscious effort and sacrifice. Slowly but surely, as people invested more resources and energy into farming, their relationship with their plants intensified too: a successful harvest brought joy, a drought-stricken field, heartache.

If it seems a bit far-fetched to describe farming in such intimate terms, consider the account archeologist Gilbert

L. Wilson published in 1917 of the Hidasta women who farmed the floodplains of the Missouri River in what is now North Dakota. For centuries, these women turned that rich alluvial land with hoes of buffalo bone and deer antlers. They offered each plant the devoted care they bestowed on their children so that it would grow strong and tall. They sang to their corn because they believed that the corn listened and thrived on recognition. Crazy talk? Well, they were doing something right. For eight hundred years they matched labor with love to grow crops that fed their tribe throughout brutal North Dakota winters.

Some Canadian scientists are just beginning to figure out what all that farming fuss was about. Susan Dudley and Amanda File of McMaster University in Ontario found that plants, like humans and animals, are capable of social recognition. Plants actually recognize other plants that are related to them, and when they see another plant as kin, they refrain from competing with it for root territory. It is not known whether plants can extend any sort of social recognition to the humans who care for them, but James Cahill of the University of Alberta and his colleagues found that they do respond to human touch. Cahill showed that when people gently stroked plants once a week some grew stronger and larger, some fell prey to insects, and others remain unaffected. Who knew? We did, once upon a time.

The more we handled and cajoled our crops the more we became emotionally rooted to the land. This is how humans learned to farm and to stay put even when their attempts met with mixed results. Farming, even today, is a gamble. The skeletons of the first farmers tell a story of degenerative joint disease, deteriorating dental health, stunted stature, anemia, and malnutrition. Still, they persevered, and the more they broke their bodies tilling and planting the fields, the less they dared leave their precious plants unattended. It was probably just a matter of time before they found themselves, like the Hidasta women, emotionally attached to their plant "offspring."

Once people settled, according to Colin Tudge, they could easily have begun to keep small numbers of wild sheep and goats. Captive baby ungulates who are fed by humans quickly come to see them as mothers to be followed anywhere. Our dogs became part of this expanding "family" as well. Now acting more like protective ewes than the wolves they had recently been, dogs began to guide herds to green pastures, reprimanding stragglers instead of attacking them. At night these canine alloparents lay down with their lambs to keep them safe from bloodthirsty predators.

Corralling these animals after dark would have kept them alive but not thriving. Even if their caretakers gave them fodder, it never would have equaled the nutritional and metabolic benefits of eight hours of night grazing. Slowly these wild animals grew smaller and weaker, which helped make them a more manageable and dependable source of meat, milk, and fertilizer during the uncertain days of early farming.

Human intervention eventually resulted in domesticated cereals and animals. These transitions were gradual, but the results were dramatic. Gone were the days of wandering the wide-open landscape. No more eating the scenery and moving on. For those who settled, the memories of a nomadic life may have been kept fresh and bittersweet in the telling of ancestral tales of days lived more in the moment. It had not been that long ago that their kin had hunted about a dozen hours a week and talked the rest. Not only had this new breed of human given up their roaming ways, they had to learn to relax in the company of hundreds and thousands of new "neighbors."

The concept of neighbor was unfathomable for millions of years. Our ancestors had always lived in small kinship bands of twenty to thirty individuals, scattered some forty miles apart. A survey of the average population of the six surviving aboriginal societies today shows that the ideal group size for the nomadic life is 174.6, while the hunter-gatherers prefer to keep their numbers around 156. The Kalahari Bushmen find it uncomfortable, even dangerous, to affiliate with more than twenty-five tribesmen at a time.

It seems that for most of our history, group size mattered, and smaller was better. Ten thousand years ago the world, which had once been plenty big enough for the 10 million or so hunter-gatherers roaming it, was quickly becoming smaller and more sedentary. It is estimated that the earliest cities were populated by up to five thousand people. Now the same number of people who had once spread out over huge landscapes were crammed onto a single acre. It seems that an expanded sense of social recognition was redefining the hunter-gatherer comfort zone.

Even after thousands of years of successfully farming, herding, and living together, there's good evidence that Neolithic people still harbored a nostalgia for the good old days. Nine thousand years ago, a group of social pioneers came together on thirty-three acres of fertile land atop the Anatolia plateau in Turkey. Here in a site called Catalhoyuk, between three thousand and eight thousand people would live in multilevel mud brick dwellings over the next fourteen hundred years. They farmed and kept animals and created the ultimate tight-knit community, so densely constructed that much of Catalhoyuk could be traversed only rooftop to rooftop.

Their dates fix them as late Neolithic people, but their art tells us they were emotionally committed to their Paleolithic past. Like their Ice Age ancestors, they too were fierce hunters who painted their walls with the images of wild animals. Some of these murals, which are remarkably preserved, show large, dangerous beasts (stags, boars, and bull aurochs) being surrounded and touched by tiny men and sometimes tiny women. There are many theories about what is going on between human and animal in these paintings. Are they records of some dominance ritual or perhaps a rite of passage that involved the group harassment of wild animals? Or are they merely wishful thinking? Are the animals in them even alive, or is this bravado part of a post-mortem victory dance? They may even be a kind of proto-ad for the virility-enhancing services of some shaman. Any of these interpretations imply that these were people who still revered the very thought and image of close contact with wild animals.

Even their cleanly plastered walls and tidy floors betray an Ice Age psyche not ready to give up the ghost. There is an eerie echo of the skeletal hunting lodges of the Ice Age Ukraine in the way the people of Catalhoyuk embedded their walls, floors, and resting surfaces with the horns, skulls, scapulae, teeth, tusks, and talons of wild animals. They kept sheep and goats penned outside the village, but in their heart and homes they kept the wild close. These look very much like people who were still trying to have it both ways.

Ian Hodder, who has directed the excavations at Catalhoyuk since 1993, thinks the architecture of the settlement may also suggest that the inhabitants were trying to maintain some sense of themselves as rugged individualists. He points out that, unlike earlier Neolithic settlements, Catalhoyuk had no communal structures. Clearly settlers must have worked together to grow crops, protect their herds, and hunt. And the community did gather together for ritual feasts, roasting as many as five bull aurochs at one celebration. The rest of their lives and identities, however, Hodder says, was defined by their homes. They shared their fiercely decorated dwellings with five to ten living relatives and many more dead ones resting under the floors and sleeping platforms. This unusually intimate burial practice strongly suggests to Hodder that these people were more committed to their clan than to their community.

Their commitment to their domestic animals and plants seems to have been qualified as well. Livestock was not kept inside the village or depicted in any art. The nature of these docile animals and the care of them seem to have been considered deeply inferior to hunting and eating wild game. Catalhoyuk's dogs, on the other hand, appear to have been stuck in some time warp between the ages. Hodder says they were domesticated, but were not pets. They were not normally eaten except for an occasional ritual meal. While they are seen in murals participating in hunts and standing close to people, there's no evidence dogs were allowed in the home. In this Neolithic village they may have been considered friends, but not best friends.

Cats seem to have been a very different matter. The most fascinating mystery of Catalhoyuk is its obsession with the leopard—a fellow hunter of the Anatolian plain. Sixty-five percent of all mural paintings and 35 percent of all the wall sculptures are of leopards. People are often depicted in paintings and clay figures as wearing leopard skins or materials painted in leopard-like patterns. One of the most stunning clay figurines found by James Mellaart, who first excavated Catalhoyuk in the 1950s and 1960s, is of a naked woman sitting between two leopards resting her hands on their heads. In other artwork people are shown riding and tending leopards, leading Ian Hodder to consider the possibility that some people at Catalhoyuk might have tamed this very independent and shy animal—or again, maybe they just wished they could.

Certainly the big cats, and the leopard in particular, were very special to this community. Although it was common Catalhoyuk practice when abandoning a house to remove the real skulls and sculpted heads of animals, the leopard reliefs were left behind with heads still intact. And yet despite the overwhelming artistic presence of the leopard in everyday Catalhoyukian life, no actual bones of this animal—or of any of the large cats of the region such as cheetahs and lions—have been uncovered at the site. This is in stark contrast to the 24,190 animal bones that have been identified. Ian Hodder believes that this animal held such symbolic sway over the psyche of Catalhoyuk that it may have been taboo for anyone to bring their bones and meat into the settlement.

All the art, rituals, and taboos practiced by the settlers at Catalhoyuk could not keep their world wild once the dynamic of settled life and domestication had taken hold. By the sixth millennium B.C., Catalhoyuk had fully entered the Neolithic Age. The houses built during this time are no longer monuments to the ancients and their old ways. The wild animal wall art is gone and so are the dead relatives. Clay pots with geometric decoration become the new household essential for storing, preparing, and cooking the fruits of the tended fields. In the feasting halls, the aurochs bull's bones are replaced by those of domestic cattle—a sure sign nothing would ever again be the same. For almost a thousand years, the people of Catalhoyuk had measured their worth against the deadly might of the aurochs. Now something amazing had happened, and this terrible beast had become the animal their artists had envisioned—one pliant and subordinate to their power. But how?

Attracted to the cultivated fields, the aurochs and other wild animals adjusted their free-roaming ways to stay within the sphere of this new human—the farmer. Even as Catalhoyuk's art and ritual celebrated the power of the aurochs, its farmers were beginning to see it in a new light. The aurochs was becoming a *pest*. But it made no sense to try to scare off a herd of wild cattle protected by a bull who stood six feet tall at the shoulder, with horns spanning eight feet. And even if they could kill them all, they couldn't eat all that meat before it spoiled. It must have presented a real spiritual dilemma for these farmer-hunters, especially coming at a time when crops were becoming more precious and wild bulls were becoming scarcer. For the first time, humans needed to conceive of a new, more cooperative relationship with the aurochs.

This was a task decreed impossible by no less than Julius Caesar when he and his legions encountered a wild aurochs around 58 B.C. on a military campaign in Germany. (The last wild aurochs is believed to have been killed in Poland in 1627.) In *Animals in Roman Life and Art*, classicist J. M. C. Toynbee tells us Caesar found this adversary to be slightly smaller than an elephant, strong and swift, and "showing no mercy to either man nor beast that came in its way. It had to be caught in a pit and was impossible to tame." Caesar must have been so busy consolidating his empire that he failed to notice something about the aurochs that his sharp-eyed ancestors had. Aurochs, like all hooved animals, like the taste of human urine.

Some alert early hunters or artists may have figured out that they could manipulate the aurochs's movements with the strategic placement of their urine. The person who first figured out that it was the salt in the urine that aurochs craved discovered not only a simple way to keep aurochs off the crops, but the way to harness the living power of this animal. Fredrick Zuener says that "the use of salt in binding animals to man is widespread and may have underlain the initial taming of . . . cattle and sheep." He also quotes mountain sheep expert Valarius Geist, who observed that wild sheep would accept a non-threatening human's as a two-legged salt-lick."

The allure of crops may have enticed the dreaded aurochs to take the first step toward humans, but it was human genius that noticed that change and met it with salt around 6000 B.C. It had taken the aurochs almost a thousand years longer than the sheep and the goat to "turn loose," as Ray Hunt would say, but finally some special humans had aurochs eating out of their hands. We can only guess what emotions this remarkable show of trust and submission stirred in these most reluctant Neolithic hearts.

Some people with a burgeoning knack for this sort of thing took on the care, the feeding, the grooming, and the protection of this awesome animal and in doing so, tripped a developmental switch that would turn their wild prize into a cow. Like the wild sheep and the goats before them, selective breeding and a limited diet began to reduce the aurochs's bulk and ferocity. All the while, increased human tactile contact was stimulating the cattle's oxytocin system, gentling it further and turning it into an animal like none they had ever known.

The historian Lewis Mumford sees domestication as a process of "gentling, nurturing, and breeding," and he's adamant that we not lose sight of the essential role women must have played in the Age of Domestication:

> woman's capacity for tenderness and love must have played a dominating part. . . . House and village, eventually the town itself, are women writ large . . . Security, receptivity, enclosure, nurture these functions belong to woman.

As we now know, these functions also belong to oxytocin, and a woman's estrogen enhances her oxytocin system. We also now recognize that this molecule has powerful taming and bonding effects that can cross the species barrier. This suggests women would have been biologically and intuitively qualified to make a significant contribution to the domestication of animals.

Ian Hodder says the mythologies that arose from the earliest settlements in the Levant, the Fertile Crescent, and the Nile Delta all tell of mother goddesses with the power to tame nature. James Mellaart believed that the statue he found in a grain bin in Catalhoyuk of the woman sitting between two leopards was just such a goddess. Did she tame those leopards with her supernatural powers or had they fallen under the remarkable but decidedly natural spell of her oxytocin?

ELEVEN

Made for Each Other

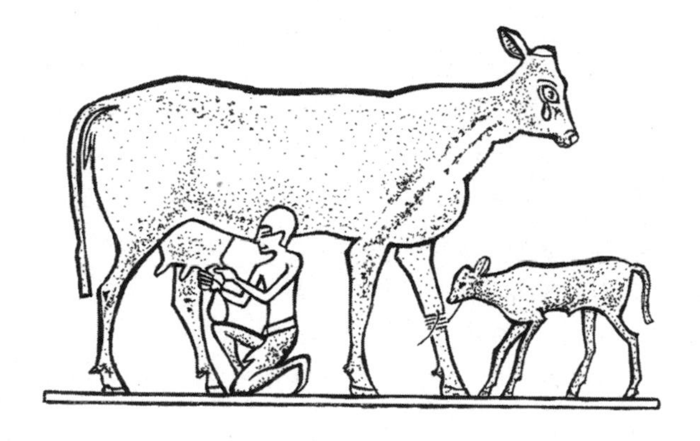

The Neolithic lifestyle also had profound effects on the breeding and feeding pattern of the newly domesticated humans. By 3500 B.C., skeletons found in cemeteries of ancient settlements such as Abydos in Egypt show rampant congenital abnormalities that may have been caused by environmental hazards or inbreeding. Bones of teenage skeletons show wear and tear from the stress of building cities and hauling harvests. Crowded neighborhoods were repeatedly thinned by the diseases and parasites humans picked up from sleeping and eating with their new animal companions. Caring for plants and animals definitely had its downside.

The new social stress of coping with so many people in such close quarters also weighed heavily on the first settlers. Hunting and gathering had required a limited range of talents and abilities that were shared fairly evenly throughout the group. Now the world needed more brawn than bravery; ditch diggers for irrigation, plowmen to break the ground, cutters to harvest the grain, and grinders to make it edible. And animals needed someone gentle enough to milk a goat but firm enough to keep the cattle in line. With the emergence of an elite ruling class, the scramble for social status and reputation took on a new urgency.

It's hard to put a value on the social bump gained by those who figured out the most effective way to behave with plants and animals. In Chapter 8 we saw how, even today, the trust of a wild animal can bolster a person's reputation and social worth. Two recent fMRI studies, one from Japan's National Institute for Psychological Sciences and the other at the National Institute of Mental Health in Bethesda, Maryland, showed that our brains appraise social status and reputation in the same areas that respond to monetary rewards. In other words, our reputations are neurologically worth their weight in gold, and if being good with animals and plants improved social standing in Neolithic society, people with that knack would have been rewarded in the newly emerging hierarchies.

Even for the lowly plowman, there was the satisfaction—when it all worked—of creativity. In ancient Egypt the garden was already being appreciated for more than the food it produced. Walking in gardens was a therapy prescribed by ancient Egyptian doctors to soothe their disturbed patients. That palliative and socializing effect of tending plants was rediscovered in the eighteenth century and remains potent in modern times. In the 1960s Charles Lewis used it to ignite a sense of community in the midst of poverty and rubble. He helped clear patches of New York City's Spanish Harlem to create neighborhood gardens. In precincts rife with crime, the police were amazed when these gardens were respected and allowed to grow. They were even more amazed to learn the vandals had become guardians of these tiny Edens.

In Chicago's slums, Lewis noted that the nurturing instinct promoted by gardening extended beyond the plots to the buildings that surrounded the garden, which were freshly painted and cleaned. He concluded that gardens make good neighbors. Growing beautiful and delicious things where nothing existed before brings people together and gives them a sense of control in their lives.

Caring for animals helped our ancestors accept the mind-boggling realities of urban living in more ways than anyone could ever have guessed. As the first farmers gently stroked their animals, they raised the oxytocin levels of these captured creatures, making the animals' brains, bodies, and behavior calmer and more cooperative. The warmth and acceptance felt in that touch also produced a reciprocal release of oxytocin in the keepers. In the end, both humans and animals grew more content. Just as caring for a baby releases the oxytocin that helps mothers relax into a more sedentary, repetitive lifestyle, so the nurturing aspects of domestication appear to have released a similar oxytocin effect on most of humanity. As we accepted our role as caregiver and kin to a growing range of people, plants, and animals, our daily nurturing tasks became the sorts of activities that could keep oxytocin flowing through our new extended families. The emerging sense of well-being could have helped us settle down and face another day together.

Oxytocin's power to alleviate both physical and psychic distress must have been a help during these challenging times. As we saw in Chapter 6, oxytocin released by human contact or through injection can increase a rat's tolerance for pain. It also helps heal their wounds three times more quickly. Oxytocin released by positive social interactions with humans or animals can also bolster our immune system and protect us from infection. Oxytocin can even prevent the massive organ damage caused by sepsis that often leads to death. So, despite the hard luck story told by the bones of the Neolithic settlers, this social experiment seems to have offered hidden rewards that made the entire enterprise viable.

The difficulty and thrill of the first attempts to change the very nature of nature must have been as powerful as an

Apollo launch. The first men or women to venture into these forbidden flight zones were the test pilots of their day. It would have taken a similar combination of skill and daring to milk an aurochs. Juliet Clutton-Brock, an expert on the history of domestication, describes just how tricky it is, even now, to coax a feral cow to let down her milk.

> The cow must be quite relaxed and totally familiar with the milker, her calf must be present, or a substitute she identifies with the calf, and it is often necessary to stimulate the genital area before the milk ejection reflex will allow secretion.

It would be another fifteen hundred years before human and aurochs would reach that level of trust and social recognition, but that outcome was inevitable once the Neolithic settlers in North Africa and Mesopotamia began to selectively breed their captive herds. They could only have paired the most manageable aurochs and in doing so produced calves that acted and looked different. Just as Dmitry Belyaev would rediscover thousands of years later, trying to breed out aggression, whether in an ox or a fox, can change the body as well as the mind. The artistic eye of the cave painter gave way to the breeder's scrutiny, producing sheep, goats, and cattle with signature colors, horns, shapes, and sizes. In the end, these Neolithic animal artists so altered the conformation and nature of the beast that Julius Caesar would one day fail to see any family resemblance between his cattle and the wild ox that stood before him.

Humans had gone from hunting animals to creating them. The intimacy of breeding and birthing cattle—acts of animal husbandry and midwifery that are known to produce oxytocin—left us emotionally invested in their identities and welfare. In *The Inevitable Bond*, animal behaviorists Paul Hemsworth, John Barnett, and Graham Coleman explain that during such hormonally charged events, humans aren't the only ones that become socially susceptible.

> Since the parturient cow is extremely sensitive to her environment, the close presence of humans at this time might result in the cow forming an attachment to humans that may be similar to, but less specific and weaker than, that which occurs between the cow and her newborn calf.

Still, after almost three thousand years of selectively breeding more tractable cattle and helping deliver them into the world, there is nothing in the archeological record to indicate that anyone had yet dared to test this human-cow bond by reaching under the animal and milking it.

The earliest definitive evidence that this delicate contact had finally been made comes from five-thousand-year-old cow milk residue left in a bowl in an Egyptian tomb. Other tomb art clearly illustrates how Egyptians, using gentle, intimate touch, neurochemically convinced cows they were worthy of their milk. One painting shows a child suckling a cow alongside a calf while a man massages the cow's vagina. In another scene, painted on a coffin, we see a kneeling man milking a cow into a vessel while her calf stands tethered to her front leg. The man is smiling and the cow has a tear rolling down its cheek. The artist who created this scene may have thought that the cow was crying, but the tears are more likely a sign that she was experiencing a flush of oxytocin, released by human touch.

After a long, dangerous journey of cautious approach, and gentle, rhythmic interaction, the cow had finally relaxed enough to let her milk—and tears—flow.

Did our domesticating ancestors experience such oxytocin highs from their intimate involvement with the natural world? When the care of livestock imposed famine, disease, and despair on the first farmers, did pleasant sensations help tip the scales in favor of this more sedentary form of pastoralism? Thomas Hardy's novel *Tess of the d'Urbervilles* describes the Zen-like state that milking inspired in humans and animals thousands of years later:

> All of the men and some of the women, when milking, dug their foreheads into the cows and gazed into the pail. But a few, mainly the younger ones, rested their heads sideways. This was Tess Durbeyfield's habit, her temple pressing the milcher's flank, her eyes fixed on the far end of the meadow with the quiet of one lost in meditation.

Tess's warm and rhythmic act of "suckling" has lulled her into a meditative zone often experienced by oxytocin-rich nursing women. In fact, the entire enterprise of animal husbandry inadvertently created a potent and renewable oxytocin feedback system with profound consequences. Not only did it produce the more pleasant side effects mentioned above, but it made the downside of domestication more tolerable.

Despite the extraordinary talents of these animal-bond specialists, most of the wild animals they managed to approach and interact with failed to relax into full domestication. We know from artwork created between 2700 and 2000 B.C. that the Egyptians had developed a system of quid pro quo between gazelles, cranes, cheetahs, leopards, elephants, asses, antelopes, pigeons, and possibly giraffes. These exotic creatures ate from their hands. Egyptians must have had a gentle touch, because their art shows several different kinds of antelopes wearing collars, which means they were able to tame and perhaps even breed this ultraskittish animal. Other paintings indicate they were not just gentle with animals but also fearless. There is a tomb painting showing a hyena being force-fed—an amazing interaction, given the power of a hyena's jaws. In another unlikely agreement, sacred crocodiles (bedecked in necklaces and bracelets) came to the call of priests. Perhaps the most spectacular testimony to the desire and ability of the Egyptians to bond with wild animals was found in 2004 in a grave in northern Egypt. Archeologists discovered the undisturbed skeleton of a giant male lion, one of the largest ever recorded. Its teeth indicate it had been kept in captivity and died a natural death.

The history of domestication is littered with such false starts. Even with all the animal wisdom in their legends and song lines, Native Americans and Aboriginal Australians couldn't or didn't coerce their creatures into domestication. White explorers were amazed to find that the Indians of North America made pets of bison, moose, and wolves. Their women even breast-fed bear cubs. But despite this intimacy, Jared Diamond notes in *The Third Chimpanzee*, "No Native American or Australian mammal ever pulled a plough, cart, or war chariot, gave milk, or bore a rider."

Why? Because, Diamond explains, taming is only the first step toward domestication. Full-blown domestication is a demanding process with exacting requirements. So strict are the standards of this exclusive evolutionary club that only fourteen large (over 100 pounds) mammals qualified for membership. That's 14 out of 148 possible large herbivores that proved calm enough to handle, willing to mate in captivity, and, most important, having a disposition to submit to a dominant individual, namely, us. In the course of charming dozens of wild creatures, the animal experts of the Neolithic stumbled upon the rare creatures that could be convinced to merge into human society.

This is the transfer of power we saw in the wolf-human and aurochs-human encounters. Zoologists call it the "assimilation tendency" or "imprinting." These terms refer to an animal's ability to perceive a human as a member of its own species—as the dominant member or even its mother. Again, these cases of mistaken identity are made possible by oxytocin's ability to enhance the certain kind of associative learning and memory that creates social recognition. Its primary purpose is to bond mother and offspring, but because this attraction is a by-product of a hormone found in all mammals, this effect has had some critical interspecies consequences as well. The cow came to recognize a human as its calf (when milking), its mate (when being stimulated for breeding), or the herd leader (when grazing). The humans who allowed those assumptions to be fostered also found themselves inclined to accept these new identities and the awesome responsibilities they implied. It takes a very special kind of person to devote his or her life to the care and feeding of animals (no matter what the payoff), and those humans who first entered into these interspecies bargains found themselves changed by the animals who looked at them differently.

Egyptians seem to have been naturals at making friends with animals. Even when animals were too wild to be domesticated, the Egyptians made them pets. The one creature that moved most gracefully between these worlds was the one the Egyptians loved best of all—the cat.

Wild cats probably wandered into early human settlements in pursuit of the mice and rats attracted by harvests. The arrival of the cat would have nicely checked the rising rodent population, sparing struggling farmers wholesale raiding and a variety of diseases. In 2007, a team of French archeologists were surprised to find the remains of a male cat carefully laid to rest with a human in a grave on the island of Cyprus. The burial, decorated with shells, polished stones, and other tributes, was ninety-five hundred years old. This means that even in early Neolithic times, some humans recognized the cat as something more than a better mousetrap.

But it would be another four thousand years before the cat really found its best friend. During that long time, cats and humans seem to have been content to keep their distance until the Egyptians invited cats into their homes. The cat seems to have accepted the hospitality on its own terms—the same terms cat owners continue to live by today. There was something about the cat that gripped the Egyptian imagination and soul. They painted them, sculpted them, and tamed them.

As long ago as the early 1980s, researchers showed that long-term cat ownership could improve a person's sense of well-being. Compared to non–cat owners, those who lived with cats for at least one year showed significant cardiac and blood sugar–level improvements. Perhaps the Egyptians also felt that the cat brought them peace of body and mind. That may help to explain why they loved this animal just the way it was and did not try to breed a better cat.

The early pioneers of domestication found themselves in awe of animals they could touch. The Egyptians touched many, many animals and were so touched by the experience that they felt compelled to create a pantheon of cults worshiping species as diverse as the hawk, the baboon, the hyena, and the ibis. By 950 B.C. the Egyptians were in full-blown worship of the cat. The cat goddess Bastet was idolized as the manifestation of the original Egyptian mother goddess. Of all the animal deities the Egyptians bowed to, it was Bastet who drew the most—700,000 pilgrims—to her annual festival each Spring.

As in Catalhoyuk thousands of years before, the Egyptian cat had achieved special status, protected by taboos. It was absolutely forbidden to kill an Egyptian cat. When a family's cat died, all members of the household shaved their eyebrows in mourning. No expense seems to have been spared to ensure a cat's afterlife. Cats received elaborate burial rights including full-scale mummification—time and expense be damned. One excavated tomb yielded over nineteen tons of mummified cats.

The Egyptians were unique in the scale and scope with which they integrated the animal world into every aspect of their society. Kenneth Clark in *Animals and Men* says the emotional investment they made with animals was exceptional as well:

> Beyond these godlike attributes the quantity of semi-sacred animals in ancient Egypt owes something to

> a state of mind that by no means always accompanies religious feeling: love. The Egyptians loved animals. This statement will be dismissed by anthropologists as sentimental modern nonsense; but it is evident that the Egyptian feeling for animals was far closer to our own than that of any other ancient people.

The depth of appreciation and emotion that animals inspired in the civilized human heart was fueled by a fully matured and engaged neocortex. The big human brain had come into full bloom, capable of complex thought and self-reflection, complete with the language to express this new level of awareness. When filtered through the new and powerful neural circuitry, the ancient hormonal drive to form emotional attachments emerged as concepts of respect, appreciation, admiration, devotion, and love. The big human brain had created a big human heart that reached out to animals and enfolded them into our culture and consciousness.

Just as there was an age of exploration and an age of reason, the span from 10,000 B.C. through 2500 B.C. can be seen as the golden age of animal bonding. Humans had stared at animals long enough. Now they moved in to touch and calm as many of them as they could. It must have been thrilling and dangerous, amazing and amusing, but none of these first tamers could have imagined just how completely these new relationships would change them, their world, and ours, forever.

Those who first discovered a way with the dog, goat, sheep, cat, pig, cattle, fowl, and horse tapped into one of the most powerful biological forces on this planet—the biology of bonding. Our chemistries, our emotions, our fates became inextricably linked. Simple, gentle touch changed how we saw these animals and ourselves. We had gone beyond relationships of convenience with animals to merge with them. The bond was as intense as it was surprising—a blessing and, of course, a curse.

Domestication was our first foray into biotechnology. It may have been unleashed by natural forces, but humans were quick to exploit the advantages posed by such changeable times. Imagine the kinds of debates that may have taken place between the die-hard hunters and the first farmers. Did the nomadic hunters warn the sedentary farmers that they were entering into an unholy alliance with the Other? That they would end up slaves to their charges?

Today we are on the verge of reaching out and touching our very DNA, and many worry that this close encounter will have consequences equally profound and unintended. Francis Fukuyama, an expert on bioethics, cautions that culling or harnessing genetic material will give societies the power to control the behavior of their citizens, a power that will "change our understanding of human personality and identity . . . upend existing social hierarchies and affect the rate of intellectual, material and political progress." Such powers, he warns, "will affect the nature of global politics."

These are the same biological, psychological, and social mutations that some hunter-gatherers may have railed against ten thousand years ago. Keeping, culling, and harnessing animals would transform them and the world they had always known. Only a few would resist: the pastoralists. In the next chapter, we will meet the people who opted out of farming to pursue greener pastures with their herds. We will see how a desire for a life less confining would weave these restless souls into the tightest and most endangered of all the human-animal bonds.

TWELVE

The Survivors

To the south of Egypt, along the cataracts of the Upper Nile, the great Nubian empire also made its peace with animals. The bountiful floodplains of the Lower Nile that kept the Egyptians and their animals fat and happy were unknown to the Nubians and their herds. The Nile that roared through their kingdom was not navigable or nurturing, and the spray of its cascading waters quickly evaporated over the sands of the hottest and driest desert on earth. Still, from tiny oases along the river, the Nubians built one of the longest-lived civilizations in the history of the world.

And one of the most mysterious. The great Nubian cities were hidden away in almost complete isolation. In the ancient world, legends of Nubia were many, while eyewitness accounts were few. Greek and Roman historians tell of a Nubia filled with fantastic temples, bold, handsome warriors who could ride horses, and a queen who ruled from a palace on wheels drawn by the strength of twenty elephants. But it was the Egyptians who knew them best and liked them least.

For three thousand years the Egyptians and the Nubians struggled to rule the Nile. By the time the Egyptians had invented hieroglyphs, this rivalry was a thousand years old, and their scribes found the "vile" and "wretched" Nubians worthy subjects. In their gloating, however, the Egyptians immortalized the great pastoral skills of the Nubians. In 2600 B.C. the Egyptians recorded a great victory over this empire with spoils that included 7,000 Nubian men and 200,000 domestic animals. This historical footnote testifies to the Nubian commitment to animals. Only great ingenuity, sensitivity, and willpower could have transformed this land of rock and desert into a habitat that supported such a massive interspecies enterprise.

It is believed that the wild aurochs and its domesticated descendants became, first and foremost, an object of adoration in early human settlements throughout the Middle East, including Nubia. Sacred bulls were specially bred to be worshiped in life, then sacrificed to enlist the help of the gods or appease their vengeance. The strength, bravery, and virility of these mighty animals made them more worthy companions and messengers to the gods than mere humans. As these people became more sophisticated in the care and breeding of cattle, they began to appreciate their considerable earthly powers as draft animals and providers of protein. Nubian artists celebrated their nation's mastery of animal husbandry in their portraits of hornless and long-horned cattle of varying colors. One tomb painting shows a chariot drawn by a pair of highly decorated piebald bulls fitted with mouth bits and driven like horses. The animal-adoring Egyptians must have been thrilled to have captured vast herds of grade A Nubian livestock.

By the third millennium B.C., cattle had become the most valuable animal under human control. Their slaughter and consumption were strictly regulated by high priests and limited to sacrificial occasions. Milk, not meat, became an important part of the diet in many Neolithic societies. And as a draft animal, the ox was incomparable. Along the Nile and throughout the early farming communities in the Middle and Far East, cattle had become critical to human happiness in this world and beyond.

The status that ancient cattle bestowed on their owners would have inspired the best and the brightest to devote themselves to understanding and nurturing these prized animals. How the Nubians created their remarkable human-animal bonds will never be fully known. Much of the Nubian tale is prehistoric. What we know about the Nubians comes from unflattering accounts told by their more literate enemies. The Nubians did not write down their stories until 170 B.C., and these tales are locked in a cursive script that has yet to be fully deciphered. Much of what's left of this empire now sits at the bottom of Lake Nasser, which was created by the Aswan dam in the late 1960s. Upper Nubia, now Sudan, remains obscured by intermittent civil wars and the enduring brutal terrain. Today what we can still see of Nubia is glimpsed in fashion and furniture, but most clearly revealed in the cattle and their keepers who descended from this noble stock.

The Nubians and the Egyptians left an enduring legacy in the great cattle cultures that have fanned out across Africa from the Nile to the Atlantic. Their clear imprint can be seen in the cattle-conscious societies that exist in sub-Saharan Africa today. In the 1930s Sir Edward Evans-Pritchard told the world about the Nuer people, who live in what remains of the Nubian kingdom. Evans-Pritchard found lives that were still described and decided by cattle. His anthropological survey of the Nuer and their cattle documented a two-species civilization built on "mutual parasitism."

Nuer cattle enjoyed a status almost equal to their human keepers. Grooming and decorating them was a source of pleasure and pride to their owners, who created their own names from the physical attributes of their favorite animals. Among the inhabitants of these ancient Nubian lands, Evans-Pritchard found men and women who were still known and judged by the cattle company they kept. The Nuer society reflects the major ego shift that swept over humans shortly after their domestication experiments began. Hunting animals had certainly left its mark on human culture and consciousness, but we were no longer staring down animals. Now we were staring deep into their eyes. What humans found looking back would be no match for the stout heart of the hunter. Here's how anthropologist Yi-Fu Tuan put it:

> Hunters may respect but they do not love the game they hunt. By contrast, pastoralists spend much time taking care of their livestock and are known to show strong affection toward it. . . . Cattle, to the Nuer, are not just a resource to be used. Far from it. They love their cattle.

This remarkable change of heart is what can happen when people and animals live together. As our commitment deepened to caring for whole herds of animals, our potential exposure to oxytocin's many bonding effects was exponentially multiplied. But it wasn't just the numbers that overwhelmed us; it was the nature of our interactions that finally merged us. Here, for instance, Evans-Pritchard gives us an example of the kind of mutual parasitism that can cause humans and animals to fall in love with each other.

> When [a young man's] ox comes home in the evening he pets it, rubs ashes on its back, removes ticks from its belly and scrotum, and picks adherent dung from its anus.

Gross? Of course, unless you are under the influence of oxytocin. Then, as any parent knows, it's not gross, it's not really even work—it's love. Nurturing touch releases oxytocin in the groomer, and for the cow (or the baby) the anal-genital stimulation is a powerful trigger of the hormone. Once human-animal contact became as intimate and constant as this, the emotional tradition of the hunter gave way to something quite new. In the language Evans-Pritchards uses to describe the emotions of the young Nuer men it's obvious that oxytocin has entered the equation: "No sight so fills a Nuer with contentment and pride as his oxen." Contentment and pride: the feelings that surge through any new mother.

The Nuer, like their Nubian ancestors, once carved out a Nile empire from lands dominated by another great human/cattle society—the Dinka. The Dinka, who share the Neur's prehistory, linguistic roots, and love of cattle, are the largest pastoralist society in southern Sudan. In the 1950s and 1960s, anthropologist Godfrey Lienhardt lived with the Dinka and documented how their mutual parasitism was enhanced by a mutual sense of understanding. In his classic study, *Divinity and Experience: The Religion of the Dinka*, Lienhardt tells us that the Dinka "treat cattle as though the beasts had a kind of understanding of the wishes of their human guardians." He also noted that "there is a large vocabulary of cries to use in herding and addressing cattle" and that "some beasts are more intelligent and responsive than others."

Both of these great, ancient cattle cultures have been blown apart by the civil war that has raged in the Sudan over the past fifty years, and both face the trauma of unwinding their ancient identities from the cattle they can no longer keep. Even stripped of most of their herds, the Dinka continue to honor their cattle kin through the cow dance, in which Dinka women hold their arms above their heads in imitation of their favorite long-horned cows, a vision that could have been lifted from the decorations on predynastic pottery. A Nubian tomb painting illustrates how cattle once returned this tribute. The artist shows a cow being presented as a royal gift. Atop its head sits a likeness of a human face while its two long horns are capped with models of human hands making them appear like the arms of a dancer. Cows imitating dancing humans imitating dancing cows. In the end, this Neolithic dance may be all that remains of the great love that bloomed between people and cattle along the Upper Nile during those first heady days in the age of bonding.

Colonization, war, famine, and the endless search for greener pastures have spread the Nile's cattle cultures across all of Africa. One herding society that has flourished in these hard times is the Fulani. The Fulani's origins are unknown, but they are believed be the descendants of the Egyptians. In the past thousand years their wanderings have taken them completely across sub-Saharan Africa. Many Fulani have become professionals and taken key roles in the politics of the lands in which they've settled. One group, however, has shunned modern life in favor of the rugged bush that suits the fierce hardiness of their cattle.

The Fulani live in clans made up of individual families, each tending about fifty cattle. Unlike our cows, these animals have long horns and feisty tempers, more like the wild cattle confronted by our ancestors. Like their predecessors, the Fulani manage these large, aggressive animals without the aid of barbed wire, corrals, squeeze chutes, tranquilizers, horses, or dogs—the herding conveniences essential to the modern cattle industry. Yet one Fulani herder can keep fifty hungry animals out of succulent, unfenced, cropland and on a forced march of up to twenty miles a day. How? With long sticks and a remarkable understanding of how cattle think and feel. Armed with these sticks, their animal understanding, and some incredible moxie, the Fulani actually enter into cattle society—and take over.

Every herd has two ruling animals that must be dethroned. First is the dominant bull. Fulani boys, by the age of six, must be able to intimidate these very dangerous, very large animals. Tiny boys are expected to walk up to the largest and most belligerent animal and hit it with a stick. A bull's response to such an impertinent attack will be instant and violent. Without blinking, the child must strike again and again until his very fierceness intimidates the animal into submission. The next confrontation is with the herd's dominant cow. In cattle society these females control the movement of the herd. As nomads, the Fulani are always on the move and need cattle that are willing to follow *them*. Each herder's success depends on his ability to face down the lead cow and take over her alpha role.

Moving cattle is tricky business. A herd is like an amoeba, prone to spilling out in several directions at once. Even with the help of dogs and mounted cowboys bringing up the rear, you have the constant annoyance of stragglers and the ever-present danger of a stampede. Fulani cattlemen avoid all that tiresome wandering and panic by getting out in front of the herd and *pulling* them along. By assuming the position of lead "cow," the Fulani offer their herd the only guidance they understand and respect. In this natural lead position of dominance, the Fulani herder can alert his followers with a special call. When he starts running, the whole herd immediately falls in behind, following their leader in single file or pairs. This orderly procession can be maintained all day, until the herd arrives at the next campsite. It's a trick the Fulani are rightfully famous for. Outnumbered fifty to one, they maintain complete control of their entire herd, while even a single cow from their herd can prove too much for an outsider to handle.

The Fulani attribute their way with cattle to the unflinching courage they show when challenged by beasts far more powerful than they are. But their acceptance in the dominant role of another species' society is not solely the result of brute force. This control is largely finessed through the Fulani's ancient understanding of the intentions and behaviors of their herds. The timing and severity of the herder's aggression are neither cruel nor unusual because they perfectly imitate the dominance behavior used by the cattle themselves.

The Fulani realize there is more to herding than leading animals to food and water. A truly successful herder must have cattle that not only fear and respect him but trust him too. For instance, the sting from a stick may demand respect, but it will not induce a cow to release her milk. Quite the opposite. The Fulani herders must drop their bully act if they want to be able to move among their animals to milk, groom, and doctor them when necessary. These gentle, intimate encounters require quite a different social strategy, so once again the Fulani take their cues directly from their cattle.

When cattle are not competing for social dominance or following one another, they are often licking and grooming each other. This softer side of cattle life keeps the herd healthy and clean and strengthens their social bonds. Every Fulani herder must also learn to put down his stick and his bravado long enough to understand and emulate these friendly aspects of cattle culture as well. In perfect imitation of cattle grooming behavior, the Fulani stroke their animals' heads and necks. As a special treat, they pet the upper inside of the back leg—the very spot where a mother licks her calf. The animals welcome these nurturing moments, standing transfixed, and even lick their owners in return. These simple acts of kindness physiologically release the oxytocin that then expands the level of understanding between the two species—with the human now soliciting and accepting the role of ally, even parent.

The success of the Fulani's herding methods is not without cost. Aggression, domination, assertiveness, and retaliation—the tools of their pastoral trade—have become prized virtues in their society. Their fierce reputation has spread well beyond the animal world. Fulani men advertise their bravery through the terrible scars they acquire in ritual beatings. Although a small minority, these pastoral people are given wide berth by the inhabitants of the lands they pass through. The Fulani are known to be easily provoked, and their social code demands that even minor offenses be challenged with physical confrontation. If a young Fulani man's honor is offended, he must defend it or be beaten and mocked by his own family. The Fulani are as strict with one another as they are with their cattle.

No one knows when the Fulani's struggle for dominance and acceptance in the alien world of cattle began. What we do know is that without the aid of advanced technology, they have infiltrated this foreign culture by manipulating it on its own terms. Thinking like a cow, moving like a cow, communicating in "cow" has taken its toll on the Fulani and the other societies who've linked their fates to cattle.

Dale Lott and Benjamin Hart are two ethologists who have studied what happens when humans totally immerse themselves into the society of another species. Their work with the Fulani focuses not so much on the Fulani's amazing abilities to socialize their cattle, but on the socializing effect the cattle have had on the people who tend them. These humans have, according to Lott and Hart, become hybrids whose psyche is "part and product of the behavioral properties of the cattle." The Fulani's formula of intimacy and aggression, courage and understanding, sounds like a description of how oxytocin and vasopressin promote social interactions. We know that oxytocin can flow between humans and animals, and it's possible vasopressin does as well. When the conditions are right, these two bonding agents may even be able to create a single society out of two very different species.

The tough conditions of the sub-Sahara have produced tough people who tend to like tough cattle even better than most of their own kind. Elsewhere in the burgeoning civilized world other people met other animals and came to other sorts of arrangements.

India, of course, made a bond with animals that has never been broken. The cow is still sacred there, and all life-forms are believed to be equal. India's 1 billion plus population shares a fascination and respect for animals that was learned in another time. Indians have retained the same generous spirit that once helped them weave another animal into human society—the camel.

The first great Indian civilizations grew out of an arid landscape. Rajasthan, in central India, became known as "the land of the kings," with great wealth and talent and ambition, limited only by the hostile landscape surrounding it. The camel alone moved with impunity across the hot, roadless sands, inviting visions of far-flung trade and a great mounted military future. The problem was, the camel hadn't been domesticated. Hindu tradition has it that the Lord Shiva intervened and created a special caste of humans to care for this noble but independent animal. These people would be called the Raika. It may have taken some divine intervention because the camel did not easily fold into the domestication scheme. In fact, it was one of the last large mammals to be domesticated, only reaching that status around 3000 B.C. The Raika took this task to heart. They nurtured this fragile and obstinate animal until it was a creature both tame and strong. They made a camel that can carry a thousand pounds more than an elephant—a beast fit for a warrior king.

The great Rajasthan empires and their enemies have faded away, but the desert remains a formidable foe. Camels still cross the most rugged terrain without water or imported fuels—precious and rare commodities. The stamina and stoicism that made the camel a comrade-in-arms continue to make it worthy of its modern commercial tasks. While the Raika have always bred for these qualities, they've never relied solely on nature to produce a great camel. They always have and always will tutor each calf in its social and desert duties to ensure that it is a worthy member of Raika society.

The camel's job may have changed, but the Raika's work remains virtually the same. When the herd must migrate to fresh pastures, a few men venture out with as many as one hundred animals. With a blanket, a rope, and a clay milking vessel, they journey together for hundreds of miles. The men may subsist solely on camel milk for weeks at a time. This is when both man and animal become steeped in the oxytocin experience that cements their powerful emotional bonds.

Ilse Kohler-Rollefson has been studying camel pastoralists for almost twenty years. This is how she describes the Raika's extraordinary relationship with their camels:

> the way the Raikas supervise and control the movements of hundreds of camels without any visible effort is particularly impressive. Much is done by voice, and a simple command can suffice to separate mothers and young into two different groups. . . . In the case of some camels, anybody can just walk up and milk them, while others have a close relationship to a particular herder and can be easily milked only by him. Many will not yield their milk unless their young are nearby, but some will comply when talked to in a sweet voice.

For the Raika, aggressive domination has never been the best way to "get the job done." Their success in raising the perfect camel is based on the sense of kinship they form with their animals. A camel is both dowry and best man at a Raika wedding. The Raika alone honor an absolute taboo against the slaughter of camels with a simple logic: "The camel is our best friend—why should we kill it?"

The Raika have "civilized" their animals through genetic engineering and gentle persuasion. The Fulani have conquered their herds from within. Both methods work, and both methods have left their mark on the cultures of these animals and their humans.

As we learned to make greener pastures, most people stopped following or leading herds. We bred the aggression out of our animals and buried ours under layers of civilized behavior. Our seasonal wanderings have slowed to a complete halt in front of televisions and computer screens.

Even among the Fulani, some have left the herd for civilization. Curiously, many still return to earn their scars in *sharo*—the ritualized beatings. This pull of ancient traditions learned while living with cattle has led Lott and Hart to wonder what other animal-keeping learning rules have percolated into modern life. They note that the social character of the American West is largely the result of the aggression that ranchers employed in the domination of their horses and cattle. Even modern truck-driving cowboys are still considered a "breed apart," but perhaps like the Fulani and the Raika, they are more accurately a "hybrid apart."

In their quest to discover where the Fulani end and their cattle begin, Lott and Hart found some fascinating answers, but it is the question they were left with that reaches into the past and haunts us all:

> For centuries we have studied the animals who have come to share our lives. We have asked how much and in what ways we have been their creators. Perhaps we should also ask to what extent they have been ours.

THIRTEEN

The Kids in the Coal Mine

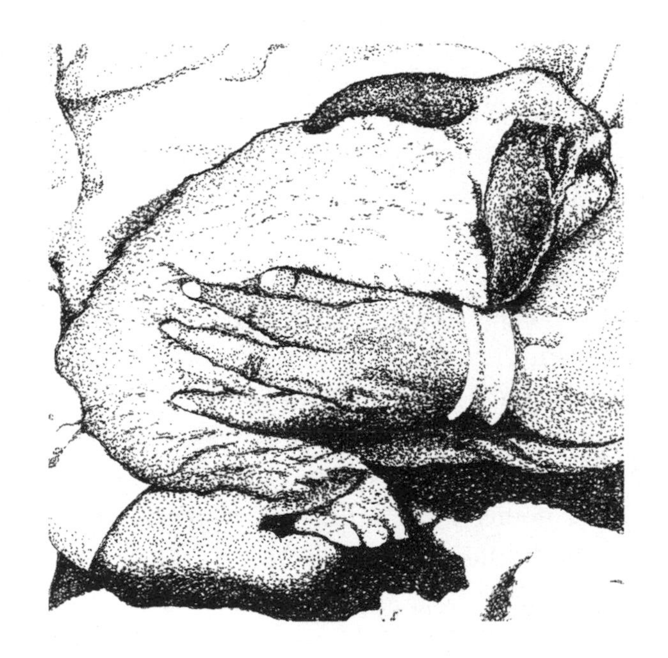

With domestication, humans and animals completed their social contract. Millions of years of approach, followed by ten thousand years of intimate interspecies interaction, yielded a relaxation that made our modern civilization possible.

As different species merged in the Neolithic period, we engaged in the kinds of sensory, tactile, and social interactions that can unleash waves of oxytocin in humans and animals. That oxytocin boost could have helped establish new inhibitory controls over the ancient and powerful urge for going. This chemistry, which was designed to convince mothers to stay put and to devote themselves to repetitive, nurturing maternal chores, seems to have gained a grip on all humanity and the animals they embraced.

Now consider that until very recently, America was a nation of farmers and herders. It may seem like a million years ago that we lived like the Raika or Fulani, but just one hundred years ago, half of America still lived and worked on farms. Now it's less than 1 percent. We are the first society to walk away en masse from the animals that humans have lived with for ten thousand years. If it is true that over millions of years animals became not just part of our lives but part of our bodies and minds, what happens when we try to live without them?

Psychiatrist Aaron Katcher says that in our abrupt shift from farm to factory, we did a lot more than just put down the plow. More critically, he says, we broke the bond with animals that had helped to make us civilized human beings. Katcher sees the fallout from this sudden interspecies divorce every day in children who are too wild to participate in polite society.

Aaron Katcher treats young patients who suffer from severe forms of attention deficit hyperactive disorder (ADHD), a mysterious developmental disorder that has become epidemic in America and a growing menace throughout the developed world. As many as 5 to 10 percent of American school kids have been diagnosed with this condition, which makes them highly impulsive and socially disruptive. ADHD eats away at a child's ability to sit still, listen, learn, and make friends—a particularly serious condition for a species as socially dependent as we are. As Katcher explains, "Kids who can't concentrate or control their behavior have a very difficult time fitting into mainstream society."

After ten thousand years, civilized behavior now comes to most of us as naturally as language. And yet social acts that we take for granted—listening to others, speaking in turn, not injuring others, not injuring ourselves—are suddenly becoming impossible for too many children. ADHD is now the most common behavioral disorder in America, and Ritalin, the drug most widely prescribed to treat it, has become a dreaded household word.

In the early 1990s Aaron Katcher was working with kids whose jumpy, uncontrollable, and aggressive behavior forced them out of their families and communities and into a residential school and treatment facility run by the Devereux Foundation in Glenmore, Pennsylvania. In this beautiful rural setting Aaron Katcher and psychiatrist Gregory Wilkens experimented with another way to way to treat kids who cannot calm down or create meaningful social connections with other people. In 1991 they designed two voluntary nature-based programs to complement the students' regular school curriculum. Fifty boys, ages nine to fifteen, all with ADHD and many with serious conduct disorders as well, were randomly assigned to either the control program—an outward bound course that included rock climbing, canoeing, and water safety—or the experimental companionable zoo program.

The zoo participants were offered an opportunity to adopt one of the zoo's small animals as a pet of their own. First, however, they had to agree to the zoo's rules of engagement: talk softly and be gentle with the animals and the other students in the program. They were also required to learn the feeding, grooming, and breeding habits of their pet in order to earn the privilege of adopting it. Embedded in the five-hour-per-week pet care sessions were lessons in spelling, math, geography, biology, and common decency.

Almost as soon as they entered the zoo building, the students' agitated and aggressive behavior began to fade away. Zoo instructors were amazed at the boys' ability to concentrate and cooperate and by the sudden enthusiasm they showed for learning. That enthusiasm was also apparent when the researchers compared the attendance to both voluntary nature programs. The first summer 93 percent of the students showed up at the zoo, while only 71 percent took part in the outward bound activities. When residential counselors and schoolteachers rated the behavior of the students from both programs, they found a significant decline in the negative symptoms of the zoo participants compared with the control group.

After six months, the outward bound students were switched to the zoo program, and their attendance improved about 20 percent. The original zoo kids were permitted to continue visiting the zoo and their pets in their free time and did so frequently. This is another indication that the animals created a strong focus for these most distracted and disturbed children.

Of the first hundred children who participated in the companionable zoo program, eighty showed an increased attention span as well as an improvement in their ability to take part in a dialogue both in the zoo and in their other classes and dormitories. Their improved self-esteem became apparent in nonverbal ways as well. Their faces and voices took on a new animation that made them more socially engaging. This thawing of expressive affect signals an internal warming—a relaxing of vigilance that is the first step back into society. When Katcher and Wilkins started the zoo program, they had high but realistic hopes. They anticipated approximately seven violent outbursts a year based on the number of incidents that occur on campus annually. To their delight and surprise, there has never been a serious incident of aggression or uncontrolled behavior in the zoo.

The zoo program was not successful because it attracted the least disturbed boys. Many of the children who have participated in the program have needed to be confined in body restraints at times when they were outside the zoo and in the less structured atmosphere of the residences. One of the boys would smash his head in a rage, and another was found one snowy winter night running in the forest half naked. For some of these very troubled children, the zoo proved to be the only society in which they felt safe and relaxed.

The zoo program was the first controlled clinical trial of an animal therapy designed to treat a disorder previously addressed only through drugs, and it was a dramatic success from day one. About seventy-five boys pass through the program each year. Zoo director Merian Waters has worked on campus for thirty-five years. She was one of the program's first instructors, and she still marvels at the animals' power to pacify. "There is one boy whose hands shake so much I don't know how he can even pick up a gerbil, but he manages and does very well with the animals." Waters says another student who is extremely talkative and hyperactive becomes so calm and quiet when holding a guinea pig that "he just doesn't seem to be the same boy." Waters has noticed that even she is not immune to the soothing effects of the zoo. Her day begins in the dorms, where she must get eight very unfocused boys up, washed, dressed, fed, and into their classes. "I'm worn out and frazzled by first period. But for me too, it all changes when I walk into the zoo and see my four-legged children."

What is it about caring for animals that can focus and gentle the most wild child? Katcher believes that, even for hyperactive children, the animals are simply too fascinating to ignore: "The animal is a constant novel stimulus that can hold their attention. Anything that directs your attention outward and stops you from thinking will calm you down." Anything, of course, that is not threatening. Pets hold all the fascination of nature without its threat of tooth and claw, and this unbeatable combination seems to be the most effective medicine. As the boys learn to trust the animals and to see that the animals trust them, they experience the joy of belonging in a way that many of them have never known.

For most of these kids their human encounters have been far less than "nonnoxious." Many of Devereux's boys have been abused, and all have known social rejection. So it's hardly surprising that they find the animals not only intriguing but also uniquely safe and uncritical. In fact, the zoo pets may be the most potent source of nonnoxious stimulation these children will ever know—and their only way back into mainstream society.

People don't have to be cruel or indifferent to rouse our defenses. Even those nearest and dearest, it seems, manage to pose a degree of competition that challenges our physical and mental territory. Deep down we seem to harbor doubts about the sincerity or willingness of others to accept us socially. That lingering—often unconscious—suspicion is a relic of our prehistoric paranoia, and it can trigger a rise in our defensive chemistry that results in an increase in blood pressure, heart rate, and anxiety.

Karen Allen, a researcher at the State University of New York School of Medicine in Buffalo, is an expert on the mental and physical changes social interactions put us through. She has found that no matter how much we may care for another person, when we are stressed even our loved ones cannot reassure us as well as our pets can. With pets present during stress tests, Allen found "tasks such as mental arithmetic and giving a speech were perceived as challenging, whereas in the presence of spouses and friends they were reported as a threat." The calming effect of animals prevailed even with her most hostile and angry subjects, who, Allen says, "had their best performances on mental arithmetic tasks and lowest cardiovascular responses when their dogs were present."

Pets seem to be particularly gifted at helping kids in the zoo program hear and even feel what their teachers and fellow students are trying to say. The nonthreatening company of animals has been shown to boost our oxytocin levels, which seems to be just what these kids really need. Studies show that children with severe ADHD have smaller amygdalas with impaired connections to their prefrontal cortexes. These two brain abnormalities leave them overly anxious and unable to control their attention or motor activity. As we will see in Chapter 14, we now have the fMRI pictures to prove that oxytocin can quiet the amygdala and allow us to see the world as a less threatening place.

In 2007, a team of neuroscientists at the National Institute of Mental Health reported that brain imaging studies of subjects with ADHD showed that their neural disconnects are a result of delayed development. It seems that the frontal cortexes of children with ADHD mature about three years slower than normal. These brain circuits that control thinking, attention, and behavior were also the last to emerge in human brain evolution. This presents the intriguing possibility that the zoo pets are releasing the oxytocin that can bridge some of these neural gaps or perhaps engage more ancient brain systems of attention that are up and running from birth. With the modulating help of oxytocin, these primal systems of attention and vigilance may produce a state of benign alertness, in which listening and learning are possible for children with ADHD.

Kerstin Uvnas-Moberg showed that repeated exposure to oxytocin could calm very, very nervous rats and help them learn vital lessons. She gave daily doses of oxytocin to rats who had been genetically engineered to be anxious. These animals were so uptight they were unable to learn that the sound of a buzzer meant they had just seconds to move away from certain areas in their cage before they received an electric shock. But after five days of oxytocin injections, she saw dramatic, long-term antistress results. Not only did the rats have all the physiologic indications of calm—such as lowered stress hormones and cardiovascular levels—their behavior was altered as well. Faced with the same challenge, these rats could now make the cognitive connection between the buzzer and the inevitable shock. Now when the signal sounded, they ran for safety. Repeated exposure to oxytocin gave them a calmer demeanor, in which they were finally able to listen and learn.

Animal behavior experts are well aware of this phenomenon. Animal trainers will tell you they dislike working with highly emotional animals because they are slow learners. Nervous cattle put in a Y-maze, in which one branch leads to a restraining device, have great difficulty learning that unpleasant consequences are associated with that path. But calmer cattle easily figure out the better option. They take their time, looking closely down both paths before choosing. Their lack of nervousness buys them time to think.

Uvnas-Moberg also showed that chronic exposure to oxytocin via breast-feeding gives new mothers the patience and extended attention span to study their babies. Students also need this kind of devoted temperament to shut out nonessential stimuli and focus on the learning tasks at hand. This is the sort of cognitive talent that anxiety and ADHD erode.

The day-in, day-out care of pets in Devereux's zoo repeatedly exposes the children to oxytocin and its long-term calming and connecting effects. And those effects can be seen not just in the zoo but in improved academic performance in classrooms across the campus.

ADHD doesn't just cause children to be poor students; it makes them socially inept as well. Fortunately, the boys who participate in the zoo program come away with improved social and scholastic aptitudes. These kids even become quite adept at making public speeches about their pets and moving easily through the audience, addressing questions about animal behavior and care.

Public speaking is a classic stressor. Yet these boys remain focused and socially engaging throughout their performance. They have the advantage, Karen Allen would note, of holding their pets while facing their public. Uvnas-Moberg would also be impressed by the nurturing attention these boys lavish on their pets. It's the same kind of fussing that releases oxytocin in breast-feeding mothers and makes them calmer and more socially desirable.

Male rodents also undergo positive social transformations after receiving injections of oxytocin. They no longer show hostility toward other males in their territory and even seek out their company. Oxytocin also helps us remember who responded to our friendly overtures so that we can build on that acceptance and form solid friendships. The instructors at the zoo have noticed similar improvements in the temperament and personalities of the boys in the program. The students are more attentive to their teachers and better able to cooperate with their fellow students. Teachers from across the campus have found that even a short trip to the zoo can help their children become more calm and socially connected. These are the Ice Age social smarts oxytocin continues to promote because social desirability is still essential to our children's survival.

There is ample behavioral and physiological evidence that oxytocin has what it takes to be the active biological ingredient in Devereux's zoo program, but the emotional evidence is equally compelling. Oxytocin is most apparent in the heartfelt bonds that form between the students and their pets. These animals not only capture their attention, they keep it. Even when the novelty of pet ownership wears off, these distracted children stay interested because they've formed an emotional attachment. For many of the zoo kids, this may be the first time they've been loved or even tolerated by another.

It's a story as old as the hills we once lived in. What starts out as an innocent visual fascination with animals turns into devotion once we hold them to our breast and feel their warmth, their fur, their trust. We become devoted to their well-being and addicted to their affection. Endless chores become effortless as they release pulses of oxytocin through our hearts and minds.

The zoo program was inspired by earlier research showing that animals have the ability to cross social and psychological barriers when humans cannot. Aaron Katcher surveyed studies which found that patients with brain dysfunctions that leave them unresponsive to other people or the environment were still able to focus their attention on animals, even hold, stroke, and hug them. Other studies showed that Alzheimer's patients who are unaware of their caretakers or hostile toward them often address these same people lovingly when an animal is present.

Friendly dogs have been shown to have a strong, positive effect on autistic children as well. Katcher was particularly intrigued by a study that compared the behavior of autistic children's interactions with their therapists when the therapist was alone or accompanied by a dog. Hours of videotape showed that children who avoided social contact with their therapist and displayed hand flapping and other autistic-like behavior were transformed by the presence of a dog. Their repetitive behavior was diminished, and they even played and talked to the animals and other humans. But all these marvelous social gains were lost when the dog was removed from the therapy sessions.

When Katcher conducted a similar study, he found that dogs can charm even the most severely autistic children. He too saw a clear dog-related increase in his autistic patients' social functioning as well as a decrease in their repetitive behavior. One young boy's verbal skills were limited to repeating words in meaningless and inappropriate ways. In his fourth therapy session with a dog, he entered the room smiling and called the dog's name. When the dog became rambunctious, the boy turned and said, quite clearly, "Sit, Buster!" This sentence was a short but sweet leap from vocabulary to language.

Katcher says that mere exposure to animals can inspire such "bursts of social competence" in even the most withdrawn and distracted humans. In the zoo program, Katcher and Wilkins upped the level of animal exposure to its most intense manifestation—pet ownership. In doing so they managed to expose the kids to a steady flow of oxytocin—the kind that can weave fleeting bursts of sociability into long-term friendship.

There is increasing scientific evidence that social competence inspired by animals is oxytocin related. The blood level of oxytocin in most autistic children is half that of normal kids. Clomipramine, the drug that autistic patients take to decrease their compulsive movements and behavior, also increases their oxytocin levels. Psychiatrist Eric Hollander, who showed that autistic patients with intravenous oxytocin improved their ability to understand emotions expressed in speech, recently conducted a study in which high-functioning adult autistic patients were given repeated doses of intranasal oxytocin. A significant number of patients had reduced repetitive behavior and showed less "cold type" social behavior and negative affect or body language.

~

Studies have also linked the impulsive behavior and scattered attention of kids with ADHD to their brain's inability to make a sufficient supply of dopamine and noradrenalin. The tons of Ritalin consumed by millions of children every year is believed to be effective because it manages to boost the dopamine levels in their brains. As we saw earlier, dopamine injections raise oxytocin levels in rats. Noradrenalin also triggers the release of oxytocin. Kids with ADHD and behavioral disorders, such as the boys in the zoo program, are often treated with antidepressant drugs designed to raise the amount of serotonin in the brain as well as antipsychotic drugs to treat their aggression and anxiety. Kerstin Uvnas-Moberg has proven that injections of SSRIs and certain antipsychotic drugs dramatically increase the secretion of oxytocin in rat brains, and she has come to believe that it is the oxytocin being released by these drugs that produces their calming and socializing effects.

In 2003 Johannes Odendaal and Roy Meintjes showed that when humans and their pet dogs get together, the oxytocin levels in the blood of both are almost doubled. They also showed that this warm, friendly contact between two species triggers an increase in beta endorphins and dopamine levels as well.

The pieces of the puzzle are falling together. Oxytocin, whether it's released by friendly human contact or various chemical agents, can make us smarter, calmer, friendlier, healthier, even more attractive. But our pets may do this best, since they seem to be particularly good at filling us with oxytocin. For this reason—and countless others—we simply couldn't have a better friend.

It's been over a decade since Katcher and Wilkins set out to bring scientific scrutiny to the body of evidence showing the great power animals have over the human heart and mind. Their zoo program was conceived to offer a nonpharmaceutical treatment for ADHD and behavioral disorders. The two men designed an environmental curative for problems they suspected were caused by profound environmental imbalances. They hoped that by putting these wild children back in touch with animals, they might give them something better than a lifetime of drugs.

It turns out that their touchy-feely therapy was chemically correct. The sensory and behavioral delight experienced while caring for animals just happens to be one of the most potent mechanisms for unleashing the brain chemicals missing in children afflicted with ADHD. Pets may not be pills, but it turns out they are very strong medicine.

When it comes to understanding the therapeutic power animals can have over humans, science has been left to play catch-up. Seven hundred years before the companionable zoo program, the duke of York recognized how horses can gentle men:

> Men are better when riding, more just and more understanding, and more alert and more at ease, and more undertaking . . . in short and long all good customs and manners cometh thereof, and the health of man and of his soul.

The duke, of course, was no scientist, but his observations were right on the mark. It may be the oldest story ever told—animals make us better people. It's just that now we are beginning to understand why. This new and ancient awareness begs the question we will explore in the next chapter: Are we, like the boys at Devereux, all suffering from oxytocin deprivation?

FOURTEEN

Oxytocin Deprivation

For thousands of years we touched animals all the time—and they touched us. If animal contact was raising our oxytocin levels, and oxytocin was even partly responsible for positive social behavior, what type of society have we created in their absence? Are the kids at Devereux—and all of us—suffering from oxytocin deprivation? To answer that question we need only look at the mental and physical changes that have crept over us since we stepped away from an animal-reliant existence.

Myrna Weismann, author of one of the most comprehensive studies of the collective consciousness of those of us born after 1955, has found that worldwide, children born in the last half century were three times more likely than their grandparents to suffer major depression at some point in their life. World wars seem to have been less traumatic to the young psyche than the peace and prosperity that followed. In Great Britain—the birthplace of the industrial revolution—the twenty-first century has produced a generation of kids with unprecedented emotional and behavioral problems. From 2000 to 2002, British children consumed 68 percent more antidepressants, stimulants, antipsychotics, tranquilizers, and antianxiety medications than ever before.

If animal deprivation leads to oxytocin deprivation, kids would not be the only ones feeling the loss. And they're not. The twentieth century made all of us more nervous, and if recent psychological studies are correct, we will soon be recasting those jittery times as the good old days. At least 16 percent of Americans will experience depression, and millions more suffer from loneliness and anxiety. The World Health Organization and World Bank estimate that depression-related health problems and suicide will be the leading cause of disease and disability in the world by 2020.

Oxytocin deficiency could well be a major factor in this dreary psychic forecast. A recent study revealed that people suffering from major depression had significantly lower levels of oxytocin. And an oxytocin deficit allows the HPA stress axis to become hyperactive—a state that exacerbates a range of mental conditions from depression to bipolar disorder.

Developmental psychologist Sharon Heller believes our hyperactive stress system is making us overly sensitive. She says as many as 15 percent of adults suffer from sensory defensive disorders that make lights too bright, noises too loud, smells overpowering, and the labels in our clothes impossibly itchy. What Heller calls SDD, most of us just chalk up to life. Of course, we mean modern life—filled with a steady stream of shocks to the nervous system and devoid of the mitigating presence of nature and animals and the oxytocin they induce. This same chemical that made the pain of childbirth bearable, the wails of the newborn tolerable, the monotony of child care a privilege, also made us all a little more tolerant and a little more resilient. Now we are annoyed to a fault.

The truth is that we are, for the first time in human history, working without a social safety net, and it's beginning to show. That amazing social experiment called domestication or civilization is over, and any contentment we eked out of our efforts to connect to animals and community has scattered—with us—to the wind.

We have become unstuck for the first time in human history. We are not beholden to family clans or limited to the society in which we were born. The care of plants and animals once caused us to settle down, learn to live together, and think of ourselves as caretakers and citizens. For twelve thousand years, we sacrificed self-interest to the care of each other, our crops, and our animals.

The industrial revolution offered the possibility of a radical departure from our ancient homesteads and state of mind. In the United States the old bonds broke down quickly. In 1920 a third of all Americans—32 million of us—still farmed the land. By 1950 that number had slipped to 23 million. Forty years later it was down to 4.6 million—less than 2 percent—and a third of those farmers didn't even live on the land they were farming. By 1993 farmers were so rare that the United States Census Bureau stopped counting. The family farm was extinct.

Of course we didn't just leave the farm; we left our families, our communities, our roots. And once we got going, we've not been able to stop. From 1970 until 1999, Americans moved 425 million times beyond their county line and another 750 million times inside the county. Meanwhile, 38 million people migrated to the United States from abroad. Transhumance, a way of life we once turned away from, is back in vogue.

Modern technology has made all this moving possible. It's given us the cars and planes that shuffle us about and the jobs to support us wherever we land. And modern technology has even offered a way to stay in "touch" with those loved ones who survive as a voice at the end of a phone line or in words spilling out on a computer screen. Is virtual contact going to be enough for a species that has always craved the real thing, or does our nervous system know the difference?

Educator Ann Alpert says the nervous system has noticed and so has she. Alpert, like all teachers in developed countries, finds herself teaching children who have grown up using computers. She says she can spot the kids who spend hours in front of a computer screen from across a room. They behave differently from kids who are less bonded to computer entertainment. Alpert says she sees a "reticence" in the computer kids whenever they approach a group of children or take similar social risks that are essential to playground politics. The computer-free kids, she says, are more gregarious and curious—again, more socially desirable.

Time spent interacting with a computer is also time spent not interacting with parents, siblings, and friends. Not only are these kids succumbing to human deprivation, many are filling that void with violence. Lots of video games are violent, and those are the ones most kids love. Left to their own devices, they turn their tireless scrutiny on virtual carnage. Even when parents limit or prohibit their children from playing these war games, there are always the twenty-four-hour news channels to give them an eyeful of real-life terror and inhumanity.

Is it any wonder so many kids have overly sensitized nervous systems? We are lucky if they are just withdrawn and jumpy. The other response to such relentless virtual trauma is to become numb to it. Our kids need more hugs; they need more supervision and guidance. They need to feel close to others and spend more time outdoors with nature and animals.

The sharp increase in autism seen in the world today may be another sign that our oxytocin system is not what it once was. Autism has gone from being a rare disorder affecting 1 in 10,000 to a frighteningly familiar condition—far more common than Down syndrome or juvenile diabetes. The latest survey in the United States found that as many as 1 in 166 kids under the age of ten may be affected with autism or one of its milder variations such as Asperger's syndrome. Undoubtedly much of this sharp increase reflects our growing awareness of the disorder and our ability to diagnose it, but that alone cannot account for autism doubling in the last decade and its being ten times what it was a generation ago.

Autism is a complicated developmental disorder that can produce the brilliance of a savant, but far more often manifests itself in uncontrollable repetitive behaviors, learning and language disabilities, and an aversion to social contact. The complexity of autism's effects is believed to be caused by as many as twenty genes behaving badly. Researchers now suspect that a defect in the oxytocin gene or the gene that regulates its receptors is a big part of the problem. Oxytocin deficiency has been linked to autism's physical behaviors as well as its social isolation.

An early indication of autism may be found in a child's failure to show interest in its mother's loving voice. Doctors are now paying close attention when a young child is nonresponsive to motherese—the special baby talk used by parents worldwide that all children should delight in. Not only do many autistic toddlers ignore their mother's ooh's and ahh's, they prefer to listen to warbles made by a computer. Their brains also do not show the typical neural activity generated when normal children hear their mother's voice. These are the verbal nonnoxious murmurings that should spark their oxytocin system. They are also the kinds of nonverbal messages that Eric Hollander (Chapter 8) helped autistic patients hear by treating them with oxytocin.

As autistic children mature, their social distress can worsen, causing them to tune out verbal and visual contact and shun physical touch. This hyperavoidance further complicates normal social maturation by inhibiting a child's ability to read the critical social information expressed in gestures, facial expressions, and tone of voice.

A crucial source of social information comes to us in the form of subtle clues visually gleaned from the expression in another person's eyes and face. As we saw, oxytocin boosts healthy people's ability to detect even faint hints of emotion expressed by the muscles around the eyes. For most autistic persons, this kind of "mind reading" is impossible because the act of gazing into the face of another is almost a torture.

Using fMRI imaging technology, Richard Davidson of the University of Wisconsin was able to look into the brains of autistic patients as they were shown pictures of faces with nonthreatening expressions. He and his colleagues saw that pleasant faces caused the kind of hyperactive response in the amygdala that a normal patient would experience when seeing a threatening expression. "Imagine walking through life and interpreting every face that looks at you as a threat, even the face of your own mother," said Davidson. Postmortem examinations of autistic brains show that the amygdala has a deficit of neurons in the area that should be rich in oxytocin receptors. When healthy men inhale oxytocin, it acts on the amygdala in ways that prompt them to stare longer at the eye region of a face and even keeps this alarm center in neutral when they look at threatening faces. These findings suggest that, by quieting the fear circuitry of the brain, oxytocin can help those with autism achieve greater social competence and comfort.

The complexity and range of autism's effects make it very difficult to treat pharmaceutically. Today, various antidepressant and antipsychotic drugs are prescribed to treat certain symptoms of autism. These drugs also happen to raise oxytocin levels. They are generally prescribed in combination with intensive therapies that involve between twenty and forty hours a week of close, nurturing social interaction with a coach. There are also tactile or sensory therapies that are designed to make these children more comfortable in their own skin and less averse to touch. These treatments use the motion of swings, pressure of weighted vests, and grooming strokes with special brushes along the child's skin. Finally, there are the animal therapies we saw in the last chapter and programs such as therapeutic horseback riding and swimming with dolphins.

Temple Grandin has fought her way out of autism's fearful grip thanks to a combination of all of the above. As a young child she was tortured by a hypersensitivity to touch and sound. Social contact was painful as well. Fear was her dominant emotion, stunting her ability to empathize and interact with others. But she was very smart, and her mother saw to it she had tutors and lots of intensive personal instruction. It helped: Temple Grandin holds a Ph.D. in animal science and is an associate professor at Colorado State University.

Grandin had to figure out the social life of humans on her own. She learned to compensate for her lack of social intelligence by clinically analyzing the paralanguage of other people. She taught herself to recognize the tone of voice, the facial expressions, and the hand gestures people use to express emotions she herself could not feel.

Fear, however, was always very real to Temple, which is why she gravitated to those who were just as scared—the cows she saw on her aunt's ranch. These were living beings she could understand. They were frightened of their own shadows, terrified of sharp noises, and scared of humans. They felt trapped. And so did she. For the first time, Temple could look at another being and feel what it was feeling. She saw herself in those animals. With these creatures, she also felt a sense of connection that she never felt for a human. And she, better than anyone, knew how to help them. Using her own fear as a guide, she devoted her life and considerable visual talents to creating more humane livestock handling equipment. Her equipment is now used in over 50 percent of all meatpacking plants in North America.

Temple still battles her autistic demons as she flies around the country giving lectures to raise consciousness about autism and the need to treat livestock humanely—even when we are butchering it. This very public and social lifestyle would take a toll on anyone. For an autistic—even a high-functioning one like Grandin—the stress can become unbearable. When this happens, Grandin prescribes the same treatment for herself that she recommends for skittish cattle: she puts herself in a squeeze chute. Grandin first saw the calming powers of a squeeze chute at her aunt's ranch. Hysterical cows became docile when placed in a narrow metal container that closed around them and held them steady. Ranch hands and vets use these contraptions when they give animals their shots or do minor surgery.

Envying the touch and the comfort those cows were feeling, young Grandin made a squeeze chute for herself.

Although human touch was caustic for Temple, she still wanted to be held. So she created a way to get the kind of hugs she needed. She made her squeeze chute of wooden boards hinged in V shape and attached to an air compressor. Once she crawled into the V she could activate the controls that closed the boards in against her, delivering the deep pressure so many autistics find comforting. In later years she would add soft padding to those boards and discover another sensation in the chute that went beyond mere calm. "The pads gave me feelings of kindness and gentleness toward other people—social feelings." This new level of nonnoxious touch delivered by the padded chute helped her experience empathy toward her own species for the first time.

Temple may have felt a new closeness to humans, but it was her cat that benefited first from her newfound understanding of "nice." In the past, Temple had freaked out the poor animal by trying to give it the hard squeezes that made her feel so relaxed. Now she was filled with a desire to make the cat sense this new wonderful feeling her chute gave her. She found the cat and stroked it gently, and for the first time, it purred. "This happened immediately after I used the soft machine for the first time. I remember the exact moment I did it." For Temple, it was a profound moment of recognition. She suddenly knew how to touch an animal in a way that would make it like her. It was an emotional revelation that would change her life.

It is easy now to see how all these interventions helped Temple compensate for the emotional and chemical limits of her autism. It also clear that autistic children stand to benefit from dedicated nurturing attention, from nonnoxious tactile simulation, and from interaction with animals. All these therapies are derived from the kinds of social contact that boost the oxytocin system.

Among the lifestyle changes that have removed oxytocin from our lives, none may end up being more influential than the way we now have our babies. A society of scattered families makes it hard to find female elders to coach us through the difficulties of childbirth. The vast majority of babies are born in hospitals and delivered by medical professionals. Childbirth has become a medical procedure.

Childbirth has always been dangerous for mother and baby. Today with the liberal use of antibiotics to protect against infection, synthetic oxytocin (pitocin) to speed up labor and placenta ejection, fetal heart monitors and ultrasounds to follow the baby's progress, and epidural analgesics to eliminate the pain, having a baby has become far safer and quicker than ever before.

In fact, speed is of the essence to the new philosophy of managed labor. A large percentage of women in labor are given pitocin to hasten the process. If the mother is not ready to deliver after approximately twelve hours, doctors strongly advise her to agree to surgical delivery. Delivery by cesarean section has become more and more common in the industrialized world, even for healthy first-time mothers with full-term pregnancies. In 2004, 29 percent of all babies born in the United States were delivered by C-section. That's a 9 percent increase since 1996.

In our fast-paced society, women may be finding it harder to relax and push. Given the risks to the health of mother and baby from prolonged labor, why wait? The answer is that the pushing and pressure required to deliver a baby is the process that gets oxytocin pulsing in our brains. And that helps establish the bond between mother and child and facilitate the flow of breast milk.

Kerstin Uvnas-Moberg has shown that women who deliver their babies vaginally have a more robust release pattern of oxytocin than women who have C-sections. They are able to hold and nurse their babies immediately, which provides a powerful oxytocin stimulus in mother and infant. Video surveillance of both groups of mothers showed that the women with the increased oxytocin spent more time touching and interacting with their babies than the C-section mothers. In surveys, vaginal birth mothers report feeling less anxious and more social than C-section mothers, and they also have a better milk flow. The stress of any surgery and the medications it requires can inhibit oxytocin production and its effects. Caesarean section surgery also eliminates the second stage of labor in which oxytocin is released. These are the sorts of things pregnant women and their physicians need to consider with great care, especially in light of the growing trend toward elective C-sections. These are surgical births performed before the mother goes into labor, for medical or nonmedical reasons.

In America, the number of women who had their first elective C-section rose nearly 44 percent from 1994 to 2001. More British women are also choosing to have their babies delivered by surgical appointment. "We are at a turning-point in obstetric thinking," declared Sara Paterson-Brown, an obstetrician and gynecologist in London. This assessment also seems to be shared by 31 percent of her colleagues in London, who reported that they would choose a Caesarian delivery for themselves—even in an uncomplicated pregnancy. This growing desire to make childbearing more efficient and painless may prove to be ill-advised as we learn more about how the oxytocin system is established and how essential it is to helping us cope with stress and form the social bonds that guide us through life.

As we slip away from the rural, low-tech lifestyle that kept our oxytocin pumping, we are more exposed to an environment that may be tampering with the little oxytocin we have left. Those who still watch wild animals closely are seeing the same social disruptions appearing in animals that have been poisoned by fertilizers, insulators, heavy metals, and preservatives. The main culprit is a class of chemicals called endocrine disrupters. They target the sexual organs affecting the production of hormones such as estrogen and testosterone—and subsequently oxytocin and vasopressin. The results are what you would expect to see when the chemistry of reproductive and social behavior is disturbed. Animals forget how they have always wooed mates, built nests, avoided predators. They even forget how to look for food.

One of the most famous endocrine disrupters is DDT. Male seagulls exposed to DDT while in the egg become sexually confused and try to mate with other males. Starlings' reproductive behavior gets skewed by DDT as well. Their appetite for sex and food, singing, and displaying is about half as strong as normal birds. Goldfish, whose graceful, mellow motions lower our blood pressure at a glance, become hyperactive when dosed with another endocrine disrupter, the widely used fertilizer atrazine.

Endocrine disrupters cause macaque monkeys to forget how to play nice. As we saw in Chapter 3, play, sex, and fighting all rely on the same physical vocabulary, which is nuanced by hormones such as vasopressin and oxytocin. It is oxytocin that breaks down inhibitions and territorial taboos and keeps play playful. It is also essential to the sexual exploration called foreplay. Fun and sexual games are as far as oxytocin will go. Without oxytocin's benevolent influence, benign intentions can be misinterpreted, offense is given and taken, play bites hurt, territories shrink, and the game is over. If testosterone and vasopressin become more dominant, play can quickly decline into fighting. Scent marking is another activity modulated by vasopressin and oxytocin. Animals use scent marking to establish social boundaries. Exposure to certain pesticides causes male mice to either mark too much or too little.

The most recent and telling study that links endocrine disrupters to oxytocin abnormalities was done by a team headed by Miles Dean Engell. He treated female pine voles—a monogamous cousin of the prairie vole—with DES (diethylstilbestrol) during pregnancy and lactation. The DES-treated pine vole mothers gave birth to babies who grew into adults showing unusual aggression toward strangers. Next he exposed pregnant and lactating female pine voles to MXC (methoxychlor), a pesticide widely used on crops that has gained popularity since DDT was removed from the market in 1972. MXC female pups grew into adults that preferred to be alone. They also had reduced oxytocin activity in their brain's cingulate cortex, a region critical to emotional behavior. The study's authors conclude that exposure to endocrine disrupters can alter brain circuitry and behavior associated with monogamy.

Humans are monogamous and are not immune to the devastating social effects of a disrupted oxytocin system.

~

If we are smack in the middle of a modern meltdown brought on by a lack of oxytocin, why aren't we all taking daily doses of the stuff? If women in labor can get pitocin, why can't the rest of us?

For oxytocin to relieve stress or social phobias, it must act on its receptors in the brain regions that regulate those things. Unfortunately, oxytocin does not easily pass through the blood-brain barrier, a dense wall of cells that protects the brain from dangerous things in our circulatory system. Still, there are a couple of pharmaceutical ways to get around the blood-brain obstacle. Repeated injections of oxytocin in high doses has been shown to affect the emotional centers in the brain, but that method of delivery is neither painless nor efficient. Nasal sprays also manage a degree of penetration. The problem is that to reach the brain with the spray, you have to inhale almost three tablespoons of the substance. Even after all that unpleasantness, the effects are short-lived.

Although oxytocin in pill form would never survive long enough in the bloodstream to enter the brain, some antidepressant and antipsychotic drugs do enter the brain and produce marked elevations of oxytocin. None of this is lost on the drug companies. "It is truly one of the body's most amazing molecules," says Robert Ring, a neuroscientist for Wyeth pharmaceuticals. His laboratory is hoping to tap into oxytocin's many therapeutic powers by developing a chemical that will improve the pattern and sensitivity of oxytocin's receptors in the brain. Ring believes a drug that boosts a person's oxytocin system could be used to treat everything from autism and schizophrenia to painful nerve disorders and depression.

Thomas Insel, who now directs the National Institute of Mental Health in Bethesda, Maryland, concurs. A pioneer in oxytocin research, he has long suspected oxytocin malfunction as the cause of autism's social abnormalities. With mounting evidence that an overexcited amygdala causes social aversion in people with autism and studies showing that oxytocin can quell this reaction, Insel believes oxytocin holds great hope for treatment of this devastating disorder. With such powerful proponents for an oxytocin drug, surely its time will come. But that process is slow. And we've only just realized how much we need our oxytocin.

It will be a long, tense time before oxytocin can be pharmacologically tamed. Fortunately there are ways we can write our own oxytocin prescription. A good marriage or stable partnership provides a variety of nonpharmaceutical ways to pass oxytocin to your mate from loving looks, comforting words, and all sorts of nurturing grooming. A recent study showed that for women, hugging is more than nice, it's oxytocin. The researchers found that the women who were hugged most by their husbands or partners had the lowest blood pressure and the highest blood concentrations of oxytocin. Sexual intercourse stimulates the genital nerves that release large amounts of oxytocin in men and women. All this indicates that loving relationships mutually release the same kinds of repeated doses of oxytocin, which has been shown to produce long-term antistress effects in rats.

Married couples appear to reduce the excitability of each other's HPA stress systems. Their immune systems are more robust, their general health is better, and they are also less likely to suffer from stress or succumb to cardiovascular disease, heart attacks, and strokes. Neurohormonally speaking, connubial bliss means never having to live in a prolonged state of sympathetic arousal.

Males, in general, are more testosterone-vasopressin dominant than females, which may be why men benefit so dramatically from an oxytocin-rich environment like marriage. The male physiology left unmitigated by the oxytocin influence of a good woman can be ravaged by chronic stress, which causes cells to age prematurely. Male rats injected with oxytocin live longer than males who are not exposed to this hormone. Husbands live longer than bachelors and divorced men.

A good spouse may lessen your chance of having a heart attack but can't eliminate the possibility. If you do have a heart attack, you may not want to depend on your better half—or any other person for that matter—to pull you through. You may want to get yourself a pet.

In a study of ninety-two patients who were hospitalized for coronary heart disease, it was not family or close friends who made a difference in their survival one year later. What really mattered was whether or not the patients had a pet. In fact, pet ownership was the most important survival factor for these subjects, second only to the severity of their heart condition. Of the 50 patients who went home to pets, only 47 survived the first year while eleven of nonpet owners died within the same period. The data also showed that it didn't matter what kind of pet the patients had, ruling out the exercise of dog walking as a mitigating factor. While marriage or other kinds of social support help insulate us against a host of stressful situations, they are rarely free of tension. Fortunately we find it much harder to hold a grudge against a pet.

This startling finding was stumbled on by a postgraduate student at the University of Maryland in the late 1970s. Erica Friedmann was part of a team investigating the survival effects of human social support on heart patients when she casually inserted a question in the survey that asked if the patient owned a pet. It was only after the long-term effects were being distilled from the data that the lead researchers in the study realized that pet ownership made the biggest difference in who survived and who didn't.

In 1995 Friedmann repeated the study with 369 patients. This time she found that dog ownership was the big survival factor in the first year following a heart attack. Of the eighty-seven subjects who owned dogs, only one died; nineteen of the non–dog owning patients did. When weighed against the other top survival factors (strength of heart, absence of diabetes, and regularity of heartbeat), owning a dog gave a heart attack victim a significantly greater chance of being alive one year later.

The discovery has inspired further investigation into the power pets have over our hearts and minds. Researcher Karen Allen's work has supported, again and again, Friedman's original finding that pets, not people, bring out the cardiological best in us. As we saw earlier, Allen showed that when people are performing a stress test, the company of friends and spouses does little or nothing to reduce their heart rate and blood pressure. The presence of their dogs, however, was most effective at keeping the subjects' hearts calm and minds clear.

Researchers from the Minnesota Stroke Institute compared the heart health of 4,435 subjects—2,235 of whom were cat owners. They found that over ten years, the cat owners had a 30 percent reduced risk of heart attack compared to the non-cat owning participants. Dog owners numbered too few in this study to offer a significant effect, but lead researcher Adnan Quershi believes they offer a similar cardio benefit to their owners

These studies bring us back to the Odendaal-Meintjes study that showed friendly interactions with our pets almost *doubles* our flow of oxytocin. We humans simply can't reach this oxytocin high by ourselves or with the best intentions of others. And we don't always have each other's best interests in mind, as in the workplace, where the oxytocin gloves are completely off. In situations where competition and territory rule, we become vulnerable to stress-related illnesses. That's when pets can be better medicine than medicine.

Karen Allen put pets to the ultimate test when she paired them with unmarried, hypertensive stockbrokers. The entire group of forty-eight stressed-out subjects was placed on an ACE-inhibitor drug called lisinopril to lower their blood pressure. The drug successfully brought their resting heart rate within the normal range. However, it was no help when they were put in stressful predicaments. Allen created such situations by administering those classic stress tests—the oral math and reading challenge. The blood pressure of all the subjects spiked despite their drug treatment.

The stockbrokers continued to take the ACE inhibitor, but half were also chosen to acquire a pet—either a dog or cat. Six months later all forty-eight subjects reconvened to undergo another round of stress tests. The medication-only group had double the stress response when compared with those who took the test with their pet in the room. The nonpet participants found the tests more "threatening," while pet participants thought they were merely "challenging." Their cooler attitude and lower blood pressure are classic oxytocin effects. (New mothers with high levels of oxytocin also stayed calm while taking stress tests.)

It is also interesting that visual contact with the animals was enough to produce this oxytocin effect. As we saw in Chapter 9, looking into the eyes of dog is enough to release oxytocin in its owner. Not surprisingly, several of the stockbrokers who were in the medication-only group went out and got a pet after seeing what a difference the mere sight of a loving pet could make on their high-pressured lives.

The link between oxytocin, heart health, and stable social relationships was made clearer in a study conducted in 2005. Researchers placed lab rabbits with heart disease in three different social settings. Some rabbits were caged individually, others were paired with strange rabbits, and some were caged with littermates. The littermates constituted the stable group. After twenty-two days the cardiac rabbits living in the stable group showed a significant increase in the amount of oxytocin produced in their hypothalamus and less atherosclerosis than those placed in the other groups. It would seem that animals can produce oxytocin and healthy hearts in their animal friends as well as their human friends.

Once upon a time, not all that long ago, people weren't depressed, they were "heartsick." Deep depression takes the song out of our voice and the spark out of our eyes, two telltale signs that oxytocin has stopped flowing. People become emotionally numb and visually dull, and in general are poor company. Friends and family can grow oxytocin-weary under this dark cloud, but our pets seem to remain resilient. Studies have shown that when depression and isolation take over a person's life, it may fall to an animal to reverse it. The mere presence of a dog has time and again cut through the isolation of the most withdrawn, uncommunicative "heartsick" patients, when our best human efforts and medications have failed.

Pets not only make us healthier and happier, but they make other people like us more. In a series of studies, college students were asked to describe the mood of people in pictures who were alone, with other people, or with a cat or dog. Their response was virtually unanimous: those with animals were perceived as being more positive. When asked their opinions about pet owners versus non–pet owners, they were united in their assessment that people who owned pets were friendlier, happier, more relaxed, and less threatening than those who didn't. Modern-day pet keeping still has the power to impress.

Pets give us an oxytocin glow that makes us feel good and makes others feel good about us. Oxytocin is what Thomas Hardy saw in Tess's eyes as she milked her cow. Clearly Tess's touch had released the cow's oxytocin and milk, but that intimate connection also stimulated Tess's oxytocin, filling her gaze "with the quiet of one lost in meditation."

Aaron Katcher has noticed the special look our faces take on when we are interacting with our pets. "The appearance of relaxation," he says, "is accentuated by the character of the smile . . . which resembles the 'Madonna' smile with which parents gaze upon sleeping infants." Practitioners of Buddhist meditation recognize this as the face achieved by a deep meditative state. The "Buddha smile" is said to open the mind and body to calm, compassion, and love—all those things that pets evoke.

The state we enter while gazing at the perfection of a contented baby or pet may be as close to meditative bliss as most Westerners will ever come. This is the state of calm receptivity that makes us open to social encounters as well as very attractive to others. It is the effect that visually sends oxytocin to those we love. In this elegant design, oxytocin helps weave the social bonds that ensure its renewal as well as ours.

Today we all live closer to the edge of social shutdown than we realize. But even in our ignorance, we sense the precipice and it makes us very nervous. Psychologist Jean Twenge's research has shown that normal children today are more anxious than children treated for psychiatric disorders fifty years ago. She predicts that our anxiety levels will remain high "until people feel safe and connected to others." Fortunately animals are masters at making us feel safe and connected, and they still pack the oxytocin punch to pull us back from the edge and keep us talking and caring. Their good company makes us better company—an ancient learning rule that continues to guide us into shelters and pet shops around the world.

Our long history with animals has left us with the clear impression that they are us—only better. They've withstood the test of time. They've become very special old friends—the rare kind who come without emotional baggage. They are the coyote that does not attack, the wolf that does not bite, the cat that purrs. They are the friend we do not leave behind.

When the poet Byron's dog Boatswain died in 1830, Byron's friend John Cam Hobhouse wrote this epitaph for his grave:

> Near this spot
> Are deposited the Remains of one
> Who possessed Beauty without Vanity,
> Strength without Insolence,
> Courage without Ferocity,
> All the Virtues of Man without his Vices.

A pet is at once baby, buddy, and caregiver. One Boatswain can put us through a rich and satisfying range of emotions making us feel alive and loved and loving, but only if we let it. A study that looked at the health benefits of marriage found oxytocin-like antistress effects only when partners were loving and supportive. It takes an open and generous heart to truly bond with another person or pet and to be blessed by it.

We have distilled 10,000 years of tactile and psychic fulfillment into the care of our pets, and now we are beginning to understand why. Our pets can be fountains of oxytocin—and so can we if we *want* to be. Someday we may have an oxytocin pill to help make our lives more livable. But even then, our best bet against the twenty-first century blues will be to keep our friends and family close, and our pets closer.

The satisfaction that washes over us as we watch our pets sleep is the ancient reminder that when all is well in their world, all is well in ours. These moments may be fleeting, but they are better than any drug. We must steal them back, no matter what the cost. Our sanity depends on it.

FIFTEEN

Just Realizing

Now that we've eliminated most animals from our lives, we are beginning to recognize how vital their presence was and still is to the quality, even quantity, of our lives. We are only just realizing that we aren't done "using" animals because we were never just using them. Whether we saw them as superior, inferior, edible, or adorable, we were committed to caring for them. It was a gargantuan task, a lifetime and lifestyle of nurturing chores that left a psychic and physical impression that we cannot dismiss lightly.

These were never merely pragmatic relationships. Long before animals were practical, they were fascinating. Long before we wanted to eat them or ride them, we wanted to paint them and touch them. Even the coyote and the badger don't just divvy up the hunt; they hang out together, side by side. Why? Because our biology urges us to connect with other living beings—and that's the other thing we're just realizing.

We and other mammals have a common social biology that allows us to approach, interact, and relax in each other's company. This chemistry helped us earn the trust and dependence of animals, but we surrendered our independence in the bargain. It was painful but bearable because independence is one survival strategy that social mammals really can't afford.

The relationships and emotions that grew between us and our animals were rewarding because they were so symbiotic. The humans and animals with the strongest social chemistry were attracted to each other. Once they began to interact in friendly ways, that chemistry began to flow between them, creating an interspecies culture based on cooperation and contentment. The animals that entered into this social contract with us became the most successful species on earth, while their wild progenitors were destroyed or became endangered. The humans who opted to care for both plants and animals were those who thrived as well. The tiny populations of humans who remained pastoralists or hunter-gatherers have not inherited the earth.

The very act of taming plants and animals tamed us too. Trainers who work with tigers and lions and other large, dangerous carnivores will tell you that the most trustworthy are those who were raised by human hand. That personal touch also arouses strong feelings of attachment and happiness in the human "parent." You simply cannot change the heart of an animal without changing your own in the process. This is what was going on while we toiled and petted our way into civilization. We were really like those honey bees who, while busily gathering pollen, also happen to be creating a world full of flowers. We were so preoccupied tending to our domesticated world, we failed to see that we were sowing a social order that ruled us as much as we ruled it—or more.

So what now? Is this greatest merger in human history coming apart at the seams? Can we face the stress of modern living without the companionship and biological comfort animals have always given? Or will we realize in time that we need animals to protect our humanity? And will this realization be enough to bring about the next major shift in our attitude toward animals? It all depends, as it always has, on the quality of attention we pay to animals.

We are born with a neurochemical capacity for biophilia. What happens to that emotional spark as we mature depends on a great many things. It depends on the quality and quantity of our experiences with the natural world as we grow. There was a time when our parents and our neighbors would have been expert animal handlers. They would have been able to deliver a calf or wring a turkey's neck. It would have come naturally to them, and it would have to us as their children. They would have taught us how to touch animals and how to think about them, since our success as farmers would have depended on it.

Urban and suburban kids today are rarely comfortable handling or caring for livestock and large animals. Few parents consider such practices important or desirable, but at least 38 percent of modern parents have maintained their urge to keep a pet. The family dog has become the family farm. We may not have been able to keep the sheep, but something deep down told us to keep the shepherd.

We are not so different from those first humans who gave up the hunt. They tried to stay close to the wild animals that had defined their lives by painting them onto the walls of their new homes. Those images were an attempt to keep connected to an untamed past they weren't ready to let go of entirely. Were the pictures enough? It can take generations to suppress the frustration and resentment that arise when migrating people take on a settled way of life. Fortunately, the chemistry triggered by the care of animals may have gradually brought a degree of contentment and satisfaction to this life of endless chores and sacrifice.

Just as those ancestors were loath to give up their roaming ways, there is still a part of us that even now wants to be back on the farm. The problem is there's no farm to return to. When we left the farm, we had no idea that we were leaving it forever. We always thought we could go home again.

Farms, of course, still exist, but they're nothing like what we remember or romantically imagine a farm should be. Tess couldn't even get a job on a modern dairy farm. Now cows are lined up in long barns and connected to cold steel machines that suck their swollen teats with clinical efficiency. No more aching backs, but no more lost reverie for the milcher or the cow. The warmth and gentle rhythm of the touch and the pleasurable sense of contentment it created have been sacrificed, and an ancient biochemical connection has been lost.

There is little that is warm and fuzzy about modern farming. All the sensual and ritual aspects of farming that stimulated oxytocin's powers have been eliminated by technology. Today's good farmer must keep his banker fat and happy and his focus on interest rates and foreign markets. With 6 billion hungry mouths to feed, a modern farmer has no time to woo the milk from a cow or walk his fields behind a team of oxen and smell the earth as it turns beneath his feet. Today's megafarms are sown, plowed, and harvested from air-conditioned cabs high atop giant tractors. Gone is most of the sweat and toil—as well as the chemical reward. Oh well, you can't eat oxytocin—or can you?

Some small farmers who have managed to keep their farms working are finding out that oxytocin is a commodity too. Actually, they may never have heard the word oxytocin, but they have come to see that "farm life" has taken on a new value.

Today a small farm can be more profitable if it is opened to the public as a theme park. In 2001, approximately 63 million Americans sought out the farm experience, spawning a whole new by-product of domestication—agritainment. In Vermont this urge to return to our agricultural roots harvested $20 million; in New York the bounty was $200 million, in Hawaii, $34 million. In Missouri, one enterprising agritainer found that city folk could be quite the cash crop. Their need to spend a day on the farm brought in $10 million—about 80 percent of his farm's annual yield.

It seems that a lot of people are hungry for "the farm" itself. This growing appetite for our rural past is redefining how farmers "work the land." Farm chores now include herding busloads of urban pilgrims from the parking lot to the visitor center, through the gift shop, and finally into the barnyard where they can pet a goat, smell the hay, and see the sun streaking through the barn door. They come in search of something they never knew but still miss. They just want to make sure that whatever "it" was still exists. So they seek out these vestiges of domestication for a whiff of something older, something saner. They want Tess's peace of mind. And for a brief afternoon, they get it. It wells up as their senses pry open ancestral memories. They see it in the wide eyes of their kids and hear it in their laughter. And as they urge them to pet the lamb, a tightness in their chest fades away.

It may be hard to believe that the farm environment still has such power over us, and yet there's something about Old MacDonald that makes sense even to an infant in a New York high-rise. The names and sounds of barnyard animals are easiest to learn and the most fun. It's weird, but it works. Our connection to an agricultural sensibility is so strong that educators here and in Europe are hoping it can rebalance the culture of childhood. California's Department of Education wants to put a little of the farm back in every kid's life by encouraging school planners to design a garden and other natural habitats for recreation and educational use in every school. Their goal is to get kids back outdoors and into nature where they can stretch their legs and imaginations.

Several other states also agree that we've got to get our kids back to the garden. In North Carolina, Southern Pines Elementary School has created a four-acre playground that is also a creative learning environment. The effect this piece of land is having on the students echoes the therapeutic effects witnessed in the zoo program at the Devereux School. Southern Pines's principal, Mary Scott Harrison, says that since the parkland was created, "incidents of fights have really gone down and our test scores have improved big time": proficiency levels in math and science have risen 19 percent. This is a particularly remarkable achievement when you consider that many of the children entering this kindergarten don't even know their numbers or colors.

It's deeply encouraging to hear that a little green is enough to spark a child's sense of wonder. A child's lack of curiosity and excitement is a true danger sign, and those signs are everywhere today. The novelist Philip Roth has lamented "the narrowing of consciousness" he sees. Others warn that escape, not exploration, is the culture's new momentum. Of course that means fight/flight is perilously close to becoming the default mechanism again rather than oxytocin's more open-hearted approach to life and love. Nervous children do not make bold innovators, yet we've never needed big, exciting ideas more than we do now.

We have faced social upheaval before, and it hurts. The pioneers who came to the Great Plains from Europe were a tough lot, but even they were not immune to the devastating pain of isolation. They hunkered down against searing heat and unbroken winds in their sod huts, dotted few and far between across the flat, desolate prairie. No town, no neighbors, no mountains, rivers or streams, nothing familiar—only social and sensory deprivation.

Folklorist Roger Welsch was surprised to learn how many of these frugal sodbusters kept canaries. A granddaughter of Czechoslovakian settlers told him about finding a small box among her grandmother's few prized possessions. When she opened it she found a dead canary. Its lifeless body told Welsch how these people survived that brutal emptiness:

> The talisman in the wooden box was not just a dead bird. It had probably been her salvation, an anchor of sanity for the Bohemian girl stranded in America's wilderness. Its trill had been her only music, its feathers the only color in her life. The gentle creature was perhaps what kept her from walking eastward in the shoulder-high grass until she died, or hanged herself. . . . What had the Czech farm girl thousands of miles from her homeland told this little bird in her moments of despair and joy? What had it told her?

It told her, "I love you," "I need you." This is the message that is music to the ears of all social mammals. It is the song we strain to hear because it gives us a reason to live. It also happens to be what pets say best.

Today we too find ourselves in a strange and hostile landscape. We suffer from a modern-day form of isolation—surrounded by people and touched by none. The more disconnected we social mammals become, the weaker we get.

Loneliness doubles your chances of sickness or death. It's as bad for you as smoking, obesity, high cholesterol, and high blood pressure.

After a day of struggling for recognition in a sea of blank stares, we far too often come home to an empty apartment. We are left to lick our own wounds—a job we were never meant to do. Social grooming was a major pastime we don't even know we miss. Grooming, that special kind of repetitive touch, is the most primitive and powerful form of caregiving and communication we ever experienced. Grooming each other was essential to our survival. We needed each other to get beyond our reach and rid us of harmful parasites. Mutual touch still raises and lowers our body temperature, speeds up the healing of wounds, takes away physical and emotional pain, and often is the only way we can communicate deep feelings for which there are no words.

The ancient urge to groom is fueled by oxytocin, which may be why oxytocin-rich pets like dogs and cats are so happy to oblige us. Grooming produces oxytocin in the groomer as well as the groomee. Legend has it that Saint Roche was cured of his wounds by a licking dog who is always pictured ministering to the holy man. Licking happens to be what many dogs frantically do to a child who is about to have an epileptic seizure. Researchers learned this by interviewing the parents of children with epilepsy. Twenty percent said that the family dog seemed to know when a seizure was about to occur, and nearly half said that their dogs would begin to lick the child's face.

We don't have to look far to see oxytocin's influence in the miraculously restorative powers of pets. Rhythmic touch and good intentions can release another of oxytocin's hidden powers: it is a potent antioxidant. Antioxidants are agents of the immune system that fight the inflammation produced by "free radicals." An excess of these molecules can lead to atherosclerosis, cataracts, diabetes, and arthritis. Oxytocin can bolster our immune system and even protect us against sepsis—the inflammatory response that can lead to catastrophic organ failure. Hospital patients, especially surgical patients, are extremely susceptible to infection, and sepsis is the most common cause of morbidity and mortality in intensive care units.

This new understanding of oxytocin's curative reach further validates the role of therapy dogs in hospitals. Health care professionals have long reported that a visiting dog can reduce loneliness and the amount of pain medication a patient requires. A controlled study at the University of California–Los Angeles Medical Center showed that in just twelve minutes a visiting dog can lower anxiety, stress, and heart and lung pressure in cardiac patients. This is a strong indication that the interaction may be raising oxytocin. Now it appears that the dogs may also be bringing antioxidant protection.

No wonder pet keeping makes such a significant difference in the quality and quantity of our lives. Our pets are literally a sight for sore eyes at the end of a hard day. They are almost always happier to see us than our human companions are, rushing toward us with a greeting that cannot help but make us smile. And just like that, they are already soothing our bruised egos.

When we see a smiling face, our own smile muscles are activated involuntarily. The mirror neurons that perform this internal response can release oxytocin and its calming, socializing sensations. So, there is neurobiological truth to the old adage that when you smile the whole world smiles with you. The greetings our pets can give us have tremendous smile power. Within seconds of seeing their pleasure at our return, we are gushing over them in "motherese," universal baby talk that keeps those happy facial nerves firing. And our pets make us laugh when nothing else can, increasing the amount of air we take into our lungs, stimulating blood flow, and reducing our stress chemistry. A good laugh even relaxes our muscles for hours. Nice job.

Pets also make us feel generous, which may be their greatest contribution to our well-being. You cannot care for a pet if you only care for yourself. Pets are an antidote to narcissism and therefore a boon to the health of not just the owner but the community. No doubt this unselfish quality is what others sense in a pet owner and what makes pet owners more approachable—more socially desirable.

Not too long ago, people were ridiculed for treating pets like family. This criticism apparently did nothing to discourage the trend. The vast majority of pet owners today unashamedly admit to loving their pets more than any living creature, kin or not. The growing realization that these beloved family members make us healthier as well as happier gives us even more reason to cherish their role in our lives.

We have not just come to feel more connected to the animals who share our homes; we're also being overcome by a growing sense of commitment to the many species that didn't make the domestication cut.

The human impact on Earth is now so complete that there is almost nothing truly wild left on it. The world has become ours, and consequently all the animals in it now "belong" to us. We are no longer only responsible for the health and well-being of the handful of creatures that entered into domestication with us. Now we must reach out to all those species that got away and try to figure out how to protect and propagate them as well. Once upon a time our only wish was to tame, destroy, or capture all things wild. Our wish came true. We turned some wild animals into house pets, put others in the barnyard, and caged what we didn't hunt to extinction. We also cut down great dark forests and dense jungles and irrigated the parched deserts. Only now are we realizing that living without wild things and wild places is dangerous too.

Just as most of us will never set foot on a farm, most of will never see a wild kingdom except on television. Our lives are no longer played out in the midst of millions of animals, but our need to watch them lingers on. The number of television shows about animals has grown as the number of animals in our environment has shrunk. Sunday evening specials such as *Wild Kingdom* and *National Geographic* have given way to entire networks devoted to all animals, all the time.

The steady growth in animal programming would indicate that watching animals is still an adaptive pastime—a modern version of those murals attempting to bring wild animals into our homes. But does this virtual version of animal contact provide enough sensory stimulation to ignite the "mere exposure effect" and sustain the human-animal bond as we know it? The opportunity to be entranced by hordes of animals is a thing of the past, but it left us with a neural legacy powerful enough to be triggered by abstract representations of those intense interspecies encounters. Studies show that watching nature films lowers our adrenalin and noradrenalin levels.

How we watch animals has changed, but the desire to view them, and the way we feel when we do, appears to be remarkably conserved. We may see far fewer animals than our ancestors did, but what is left of our natural world can now be seen with a different intensity and purpose. Binoculars, telescopes, telephotography, and satellite imagery help us see farther and deeper into our landscape than our ancestors ever could. Our modern eyes may not spot the difference between two birds in flight, but today we can see their DNA and know them by it. Now we have eyes that can scan genetic landscapes and understand animal behavior from the inside out. These visual aids have helped us draw new portraits of animals for all the old reasons—to know them better, to steal their power, to help us know ourselves.

There are many ways in which technology is actually bringing us closer to animals. Computers are helping apes and dolphins "talk" to us via symbolic language programs that we can both understand. Humans are also learning to listen to animals more closely, thanks to audio equipment that can hear better than we do.

Sensitive recording equipment has detected elephant talk rumbling well below the range of human hearing. We now know that "silent" elephants can be sending subsonic sound wave messages to other elephants miles away. By playing back these recordings we can get the gist of these conversations by the way the elephants react to them. They say things like "Hi, it's us. We're coming" and "Where are you?" or "Wherefore art thou Romeo?"

The same sort of record/play back experiments have revealed language talents in vervet monkeys where none had been noticed before. These are the animals we saw in

Chapter 8 whose infant grunts turned out to be the same sort of prewords our babies make. Researchers Dorothy Cheney and Robert Seyfarth also recorded the cries adult monkeys make when they spot predators such as snakes, eagles, or leopards. When played back, the monkeys' reactions revealed that these screams were not just general alarm calls like "Yikes!" but specific warnings that included the identity of the perpetrator. For instance, when Cheney and Seyfarth played a call recorded during the approach of a leopard, the monkeys that heard the playback ran for the trees. The cry recorded when a snake was sighted caused the monkeys to stop grazing, stand tall, and scan the grasses for signs of a snake. The eagle alarm broadcast made monkeys look up and seek shelter under nearby bushes.

Thanks to our clever new tools, we are beginning to realize that man may not be the only one who gave names to all the animals and that if a monkey or an elephant speaks in the forest and we don't hear it, it doesn't mean they didn't have something to say.

What might it mean to the future of the human-animal bond if we can figure out better ways to communicate with each other? The first great animal insights turned humans into hunters, artists, and philosophers. The next level of understanding transformed us into farmers, herders, and citizens. It's hard to imagine what revelation might come next, but it's safe to say it would change us profoundly and there's no stopping it.

We humans have always been on a crash course with animals, and we're not done merging yet. In the American Midwest, humans have concocted a highly ambitious strategy for bonding with animals called the Great Ape Trust of Iowa. This time we will not be manipulating a species' gene pool until we create a creature that looks and behaves differently. This time we will try altering the life experiences of some very special animals in ways that might inspire them to build a language-based culture that can be taught and improved upon by generations to come.

The species in question is the bonobo ape. This small chimpanzee is very smart and very social. Bonobos are also the only other primate besides us to have that long genetic allele that seems to make prairie voles such devoted parents and mates. The influence of this chromosomal tuning knob may help explain why bonobos have such a gentle, egalitarian, and empathetic temperament. It may even help to explain why they are so willing and able to share a culture with us.

Susan Savage-Rumbaugh, an experimental psychologist and leader in ape-language studies, has been working with these animals for decades, and she believes bonobos have the best chance of bridging the final cognitive gap that has separated humans from animals. She's moved eight of her star pupils into a new eighteen-room home complete with kitchens they can cook in, vending machines for snacking, and computer screens to communicate with her and other researchers. The bonobos will mate at will, play in the woods surrounding their home, and decide when and if they want to receive human visitors after screening them on their video system. For the rest of their fifty-year-long lives and the lives of generations to come, they will live within this quasi-human environment. Savage-Rumbaugh hopes that mutual goodwill and technology will foster a "pan-homo culture" including language and who knows what else. If these animals can learn these wonderful things, it will not only make us rethink what an animal is, it will make us reexamine who we are.

Impossible? There must have been plenty of naysayers when the first humans floated the idea of domestication around the campfire. Even today, the notion that we may coax another large mammal into domestication seems highly unlikely. But then domestication itself was quite the fluke—a coincidence of climate, geography, and chemistry.

In the course of a few thousand years, fourteen large mammalian species that had always run free came to accept our touch and our food and gave up their wild ways. Out of thousands, fewer than twenty species were willing to join us. In the eight thousand years that have followed we've managed to find only a couple more kinds of animals calm and cooperative enough to tolerate our close company. Is this a failure of their physiology or of our imagination?

Elk are one of the breeds that surprised us. Hundreds of years ago, a European variation of elk—the red deer—was kept for food and clothing. In the late 1600s, the couriers of Karl XI of Sweden rode elk because they were faster than horses. Then we seemed to lose our way with elk, only to give it a go again at the end of the twentieth century. Hundreds of thousands of elk and their cousins, the red deer, are now being bred and bottle-fed in captivity around the world and are on their way to domestication. The journey will take hundreds, maybe thousands of generations, and even then it may just be another domestication dead end. Modern elk farmers are finding out that domestication is dangerous business. There is no fence large enough to hold big bulls in rut or small enough to keep out parasites and pathogens. Time will tell if twenty-first-century ingenuity will outsmart these ancient obstacles and manage, after all these thousands of years, to bring another large mammal into our domesticated lives.

The tame wolves who first approached humans—or allowed us to approach them—were clearly the exception to the domestication rule until perhaps now. Dolphins, some believe, are courting domestication in much the same way.

Most archeologists and zoologists assume that long before there were dogs, there must have been thousands of years of sporadic close encounters with friendly wolves. Ever since humans could write, there have been accounts of surprising visits from wild dolphins who have volunteered to play with, even rescue, humans who've entered their watery world. Pliny the Elder, the Roman natural historian, tells us that his countrymen were charmed by friendly dolphins. These animals came when called, ate from human hands, and by several accounts allowed young boys to ride on their backs. The accuracy of Pliny's stories cannot be verified, of course, but today we can see with our own eyes that dolphins show a similar curiosity and attraction to us.

Dolphins off the shores of western Australia have been seeking out human company and contact for twenty-five years. Thousands of people have come to Shark Bay to wade in the water with dolphins who eat from their hands. But the dolphins are not just coming for a free meal. They have been known to bring fish offerings to their human visitors as well.

On the other side of the world, another dolphin is testing the waters of domestication. In Ireland's Dingle Bay, there is a wild bottlenose dolphin who has chosen a life of human companionship. For over twenty years, "Fungie" has swum with the locals, piloted their ships out of port, jumped over their kayaks, and, in general, been a good dolphin friend. He associates with other wild dolphins who enter "his" bay, but when they leave, he stays. Fungie has become a beloved member of his terrestrial community, spreading joy and wealth to the locals who have benefited from the tourist money he's netted. Friendly dolphins have been a boon to the local spirit and economies of any area they've chosen to swim in.

Most of us are content to encounter dolphins and other social sea mammals in aquariums. Still, more than fifty thousand people each year find that just watching dolphins is not enough—they must touch them. To satisfy this ancient need, they travel to places like Hawaii, Florida, and the Bahamas to swim with wild or captive dolphins in "dolphin encounter programs." In one such operation, people actually ride the dolphins holding on to the dorsal fin, just as Romans said they could thousands of years ago.

A popular dive center in the Bahamas offers the "Dolphin Experience." Visitors can watch the dolphins or strap on dive gear and swim and play with the animals. What is most fascinating about this program is that its five "star" dolphins seem to be domesticated. These animals stay in closures part of the day and at night, but they also travel out to sea several times a day to swim with humans. Each time they leave their pens, they have the opportunity to swim away and never come back, but they don't.

Why? Partly because they've lost their appetite for catching live fish. They seem to prefer fish that comes from the hand of their human keepers. They also feel safe and welcomed—two magnetic forces that keep all social animals close to home. The bond these captive animals have formed with humans has withstood more than a dozen close encounters with wild dolphins in the open seas. Not only are these dolphins not seduced by their own wild kind, they will ignore them and focus only on the humans in their midst.

In nearby waters, wild dolphins have also elected to swim with humans. These animals first began to approach humans over twenty years ago. Their gregarious nature and powerful curiosity made scuba divers just too hard to resist. They approached these space invaders and found them to be not just interesting but friendly. Now they enthusiastically approach swimmers, boaters, film crews—all human comers.

Swimming with dolphins is not just for scientists and animal huggers. The U.S. Navy has been swimming with them since 1960. In their marine mammal program they've trained seventy-five dolphins who are fast in the water, can navigate with directional hearing, and see in the dark to carry out secret military missions like mine detection, sentry patrols, and rescues at sea. Besides their underwater supersenses, these animals have a powerful sense of loyalty. They work in open water with no restraint. As one of their trainers said with pride and amazement, "For the last fifteen years, we haven't lost any animals."

In 2003, navy dolphins were deployed to Iraq to search for mines in the port of Umm Qasr. They located about two dozen explosives. It was the third time the navy has sent its dolphins to war. In 1970 they were used in Vietnam, and in the first Gulf War they patrolled the waters off Bahrain. The navy insists that the animals are trained to keep a safe distance from the mines, and a presidential investigation into charges of abuse found that the survival rate of the navy dolphins was slightly higher than that of their wild relatives. Still, the use of dolphins as instruments of war is controversial. The military has a long and brutal history of putting animals in harm's way, but also has the tremendous financial resources to explore the limits of human-animal communication.

Researchers have known for some time that these animals are very social and very vocal. The navy has invested millions of dollars to see just how much language a dolphin can learn. One group of dolphins has learned fifty different hand signals and whistle cues that identify objects or request some action. Even when the hand signals are reduced to moving points of light on a video monitor, the animals get the meaning. Dolphins also comprehend that the order in which the signals are given changes their meaning. They prove this every time they "bring the ball to the basket," instead of "bringing the basket to the ball." This understanding of the relationship of word order to meaning is called syntax—a grammatical structure that spins vocabulary into language. Syntax was a good trick when humans learned it, and it's a really good trick for our dolphin friends to have picked up.

The navy selectively breeds the most social and linguistically gifted animals from its loyal ranks. In doing so, they are following the same breeding logic that made wolves into dogs and made foxes behave like dogs. There is no reason to think that this time will be much different. By selectively breeding for calm and social traits in dolphins and other sea mammals, the navy may create their own breed of "domesticated elite."

All this talk about bringing different species into closer contact with humans does not mean that we've lost our fascination for wild animals. It's just that now when we watch them, it's harder to see their overwhelming power. What we now notice is their great vulnerability to us. We are the great herd that dominates the landscape now, and it is they who must cower on the fringes. This awareness has changed our hearts and priorities. Our obsession to tame has been replaced by a burning desire to help these animals stay free. Who knows what oxytocin forces we may be stirring with our new good intentions toward the wild things?

This heightened sense of charity is beginning at home in our zoos. These institutions that were originally designed to take the danger out of close animal encounters are now being redesigned and rededicated to allow animals to remain wilder. The old concrete cages are being placed by large enclosures that more accurately represent the natural and social environment of the captive animals. These islands of "wilderness" provide both the animals and their human visitors with a taste of the lost landscape they both crave. That deep yearning is demonstrated by the more than 120 million people who flock to zoos in the United States annually. This number is greater than the combined number of people that attend all the professional sporting events in a given year. Zoos, it seems, have become a critical refuge for humans.

They are also becoming the arks that keep some of the most endangered species afloat. These animals, once mere curiosities, are now being carefully bred in the hope that their offspring will one day be released back into the wild and will disperse a genetic windfall on these dwindling species. After tens of thousands of years of breeding for docility and human loyalty, zoo biologists must now find the wild genes and select for them. Conservation societies such as the Madagascar Fauna Group have proven that captive born and bred animals can make it in the wild. From 1997 through 2001, they released three sets of ruffed lemurs raised in America into a reserve in Madagascar. Not only did these animals manage to survive, but one female mated with a wild male and gave birth to twins in October 2002. These two new babies could rekindle a species that will otherwise disappear in the next hundred years.

There are other encouraging examples of species returning to the wild and making a difference. But many zoos find these experiments both expensive and risky, and some zoos are hoping to improve their odds by creating egg and sperm databases that can be used to inseminate wild populations with the most diverse and advantageous DNA. In their wildest dreams this new breed of "animal keeper" hopes someday to reconstitute species that have been lost to us.

Where our ancestors once delighted in changing the nature of wild animals through selective breeding, we now would settle for keeping our wildest species alive. But remaking animals tough enough to survive in the wild will be a fool's errand if there is no wild for them to return to. A slight 28 percent of the African continent remains forest and grassland wilderness. Memories of Papua New Guinea's jungles may be all the Foré are left with by the year 2021. Without a fantastic reversal of ambition, the human–wild animal bond may finally come to a dead end in the twenty-first century.

Our relationship with animals has progressed from curiosity to friend, to kin, and now, to God. With the delicate jab of a needle we are now able to reach into the DNA of animals. This very special new touch may bring the human-animal bond to its ultimate conclusion.

Zoos aren't the only human-animal institution to be reinvented with our modern technologies and sensibilities. Our ability to manipulate livestock genes has spawned a new kind of "pharming," and it doesn't happen in those big red barns or in green pastures along the road. "Pharms" are invisible, tucked away from public view in unmarked warehouses. No visitors will be tracking through mud to pet these animals, because there is no mud and visitors aren't welcome. Pharms don't look, smell, or even act like farms. Farms were about growing food, and the human-animal bond they spawned was an unintended consequence. Pharms, however, are intent solely on creating the most extreme human-animal connection.

Pharming chores take place in sterile operating rooms where the fertilized eggs of pigs that have been injected with a single human gene are implanted in sows and left to gestate in the hopes that they will develop into transgenic piglets whose organs are just human enough to be "harvested" and successfully transplanted into us. Pharms are also breeding transgenic goats that can make human proteins in their milk. Someday pharmers hope to be able to serve us milk rich in human molecules that can clot blood, fight cancer, promote growth, and treat AIDS.

Still, even this most futurist vision of interspecies merging is ruled by the most ancient social code. For one species to accept another, it must first see it as kin. If a human body is to bond with animal tissue, it too will have to see it as friendly and familiar. Until recently, hyperacute rejection squashed our attempts to meld human and animal flesh. Now this natural taboo is being broken down one protein at a time. A leader in the biotech industry managed to keep a baboon alive ninety-nine days with a pig heart before the ape's immune system finally said no.

Apart from the sense of revulsion and ethical qualms such experiments raise, there are other reasons to think merging human and animal DNA is winding the bond way too tight. Many virologists have grave concerns about the kinds of animal viruses that could be unleashed in humans when tens of thousand of us are walking around with the organs and blood of another animal.

Biomed researchers argue that careful selection, breeding, and screening can eliminate most of the threat of interspecies viral infection, but not all. It's a huge risk, of course. AIDS is a retrovirus that is believed to have made the leap into humans from primates. Other scientists are working on synthetic organs that could offer a safer and more humane resource for the tens of thousands of people who need organ transplants and die each year from the lack of them.

In the meantime, we seem to have no trouble giving our hearts over completely to our pets. In one survey that measured the human-animal bond, the vast majority of pet owners said they not only recognize their animals as actual family members, they often considered them the *most* valued member of the family. Other studies have found that pet owners commonly feel closer to their pet than to their friends and family. A study of elderly pet owners found that 78 percent of the men and 67 percent of the women said their dogs were their only friends.

When asked why their animals hold such vaunted places in their emotional and social lives, most people will tell you it's because a pet offers unconditional love. Technically speaking, this is not true. Their love is motivated and rewarded by the same chemistry that fills our own hearts. The affection an animal can offer us may not be unconditional, but it is wildly generous and inspires in us tremendous acts of fidelity and self-sacrifice that offer their own powerful reward.

One study that looked at the health benefits of emotional generosity found that married men over sixty-five who provided emotional support to their wives or someone living outside their home were twice as likely to survive the next five years as subjects who made no contribution to the welfare of another person. Researchers concluded it was the giving that made the vital difference. Considering that pet owners contribute $35.9 billion a year to the care and feeding of their animals, it's a wonder they don't live forever!

Humans and animals are the ultimate win-win story, which is why we can't and shouldn't live without each other. Human-human interactions are becoming more and more endangered. Too often they are replaced by virtual contact that cannot begin to fill the emotional and physical void. At best, these alternatives can only create islands of calm, not the sea of oxytocin that needs to wash over all of us if we are to function as a society.

What we have that we've never had before is the science to explain why we can feel such a strong attachment to animals and why they can love us too. Our bigger, talking brains are starting to bring a different quality of attention to the animals in our world today. We are beginning to see them differently and so understand them differently. We may not see them as gods anymore, but we now know that they do have powers over us that we must honor.

The bond that unites humans and animals is so ancient and so deep that it seems unbreakable. That may prove to be true, but in these first days of the twenty-first century it is under a terrible strain. So much of the landscape and the animals that lived in it has disappeared. We are the first humans to witness mass extinctions. No wonder we're so nervous. And no wonder we find ourselves taking refuge in our parks, backyards, and pets. We may have lost most of our wilderness, but we have not lost our deep need to care for "others," and it looks like we never will.

Once upon a time, a handful of humans looked upon a horse and saw it as more than meat. They saw it as their ticket to ride into a new kind of world. This is the new awareness our fancy techno-tools are filling us with today. They are revealing the deeper meaning of animals. They are forcing us to see that animals do more than feed our hungry bellies. They provide emotional transportation that can take us out of ourselves and into the social territory that has made us the strongest animal on earth. And they still protect us from the beasts within that would eat our hearts out. Clinically speaking, animals are a homeostatic necessity. Like breathing, they can be denied for just so long.

We may exist because we think, but we live because we love. And a growing number of people are finding it easier to love animals than humans. The role of the human-animal bond in this century would seem to be a variation of the function it played in helping to make us civilized human beings. As the human-human bond that supported our great social institutions is weakened by short-term, long-distance love, we feel compelled to reach out to all animals and hold them closer than ever. Deep down inside we are hoping they will keep our hearts and minds open until our own species can settle down again and act civilized. And they will. We just have to ask them nicely.

Notes

ONE ⁓ Just Watching

2 **the ones who became social carnivores:** Deborah L. Manzolillo, "Day of the Hunter," *Times Literary Supplement*, May 7, 1999. Review of Craig B. Stanford, *The Hunting Apes: Meat Eating and the Origins of Human Behavior* (Princeton: Princeton University Press, 2001).

3 **sharp eyes and a social brain:** Carl Zimmer, "Sociable and Smart," *New York Times*, March 4, 2008.

3 **tireless scrutiny:** Paul Shepard, "On Animal Friends," in *Biophilia Hypothesis*, ed. S. R. Kellert and E. O. Wilson (Washington, DC: Island, 1993), p. 278.

3 **fostered mutual brain growth:** Harry Jerison, *Brain Size and the Evolution of Mind* (New York: Museum of Natural History, 1991), pp. 70–83.

4 **bones and skulls:** Loren Cordain, Bruce A. Watkins, and Neil J. Mann, "Fatty Acid Composition and Energy Density of Foods Available to African Hominids," in *Nutrition and Fitness: Metabolic Studies in Health and Disease*, ed. A. P. Simopoulos and K. N. Pavlou (Unionville, CT: S. Karger, 2001).

4 **several such armies:** Edward Hyams, *Animals in the Service of Man* (Philadelphia: Lippincott, 1972), pp. 114–115.

4 **Paleolithic humans numbered:** Grahame Clark, *The Stone Age Hunters* (New York: McGraw-Hill, 1967), p. 16. Also see *The Human Dawn: Timeframe* (Alexandria, VA: Time Life Books, 1990), p. 66.

5 **mirror neurons:** Giacomo Rizzolatti and Lalia Craighero, "The Mirror-Neuron System," *Annual Review of Neuroscience* 27 (2004): 169–192.

6 **translate visual information into knowledge:** Rizzolatti and Craighero, "Mirror-Neuron System," p. 172.

6 **by feeling, not thinking:** Sandra Blakeslee, "Cells That Read Minds," *New York Times*, January 10, 2006.

6 **mere exposure effect:** R. B. Zajonc, "Mere Exposure: A Gateway to the Subliminal," *Current Directions in Psychological Science* 10 (December 2001): 224–228.

7 **a member of the same species or not:** Miranda van Tilburg and Ad Vingerhoets, *Psychological Aspects of Geographical Moves* (Tilburg, Netherlands: Tilburg University Press, 1997), p. 9.

7 **an odd picture:** Jerison, *Brain Size,* p. 85.

8 **brain size doubled:** Donald Johanson and Blake Edgar, *From Lucy to Language* (New York: Simon & Schuster, 1996), pp.80–81.

9 **45 million carvings and drawings:** Michael Gibson, "Reading the Mind Before It Could Read," *New York Times*, April 21, 2002.

9 **vast majority of their images:** Stephen J. Gould, "Up Against the Wall," *Natural History,* July 1996, p. 16.

9 **a child's brain:** Peter Russell, *The Brain Book* (New York: Hawthorne, 1979), p. 23.

9 **more Neanderthal than Homo:** Tara Parker-Pope, "Coping with the Caveman in the Crib," *New York Times,* February 5, 2008.

9 **primal attention systems:** C. Chiron et al., "The Right Brain Hemisphere Is Dominant in Human Infants," *Brain* 120 (1997): 1057–1065.

9 **mimic the facial expression:** Pier Ferrari, "Social Imitation in Neonatal Monkeys," www.plosbiology.org, September 4, 2006; www.plos.org.

9 **recognition and motherese:** James P. Henry and Shelia Wang, "Effects of Early Stress on Adult Affiliative Behavior," *Psychoneuroendocrinology* 23 (1998): 863–876.

9 **keen interest in other life forms:** Lori Markson and Elizabeth Spelke, "Infants' Rapid Learning About Self-Propelled Objects," *Infancy* 9 (2006): 45–71.

10 **90 percent of the characters:** Stephen R. Kellert, "The Biological Basis for Human Values of Nature," *in Biophilia Hypothesis*, p. 52.

11 **animals take on human qualities:** Elizabeth Atwood Lawrence, "The Sacred Bee, the Filthy Pig, and the Bat Out of Hell: Animal Symbolism as Cognitive Biophilia," in *Biophilia Hypothesis*, p. 334.

12 **innate tendency to focus upon other lifelike forms:** Edward O. Wilson, *The Future of Life* (New York: Knopf, 2002).

12 **the evidence remains thin:** Edward O. Wilson, *Naturalist* (New York: Warner, 1994), p. 362.

13 **genuinely universal needs are hard to find:** Jon Turney, "Of Mites and Men," *New York Times Book Review*, February 17, 2002.

13 **Most children have a bug period:** Wilson, *Naturalist,* p. 56.

13 **guiding emotions:** Wilson, *Naturalist,* p. 11.

13 **hunter's trance:** Edward O. Wilson, *Biophilia* (Cambridge: Harvard University Press, 1984), pp. 6–7.

TWO — The Birth of the Bond

16 **human figure rarely appears:** Jean Clotter, *Cave Art*, (London: Phaidon), pp. 20–21.

16 **first identified in 1902:** Linda Rinaman, Thomas G. Sherman, and Edward M. Stricker, "Vassopressin and Oxytocin in the Central Nervous System," www.acnp.org/g4/GN401000051/Default.htm.

16 **spasms that release breast milk:** Kerstin Svennersten-Sjaunja, "The Science Behind Milk Ejection," *NMC Annual Meetings Proceedings,* 2004, pp. 215–216.

17 **powerful sense of attraction and nurturance:** Cort Pedersen and Arthur J. Prange Jr., "Induction of Maternal Behavior in Virgin Rats After Intracerebroventricular Administration of Oxytocin," *Proceedings of the National Academy of Sciences* 76 (1979): 6661–6665.

18 **Mother sheep:** K. M. Kendrick et al., "Cerebrospinal Fluid Levels of Acetycholinesterase, Monoamines, and Oxytocin During Labor, Parturition, Vagino-cervical Stimulation, Lamb Separation, and Suckling in Sheep," *Neuroendocrinology* 44 (1986): 149–156.

18 **oxytocin levels rise:** K. M. Kendrick, E. B. Keverne, and B. A. Baldwin, "Intracerebroventricular Oxytocin Stimulates Maternal Behavior in Sheep," *Neuroendocrinology* 46 (1987): 56–61.

21 **prairie voles' monogamous bonds:** L. Gavish, C. S. Carter, and L. L. Getz, "Further Evidences for Monogamy in the Prairie Vole," *Animal Behavior* 29 (1981): 955–957. C. S. Carter, A. C. Devries, and L. L. Getz, "Physiological Substrates of Mammalian Monogamy: The Prairie Vole Model," *Neuroscience and Biobehavioral Reviews* 19 (1995): 303–314.

21 **close neighbors:** Rinaman, Sherman, and Stricker, "Vassopressin and Oxytocin."

22 **all monogamous:** D. M. Witt, C. S. Carter, and T. R. Insel, "Oxytocin Receptor Binding in Female Prairie Voles: Endogenous and Exogenous Oestrodiol Stimulation," *Journal of Neuroendocrinology* 3 (1991): 155–161. Thomas R. Insel and Larry E. Shapiro, "Oxytocin Receptor Distribution Reflects Social Organization in Monogamous and Polygamous Voles," *Proceedings of the National Academy of Sciences* 89 (1992): 5981–5985.

22 **social recognition:** Isadora F. Bielsky and Larry J. Young, "Oxytocin, Vasopressin, and Social Recognition in Mammals," *Peptides* 25 (2004), 1565–1574.

24 **just how social the individual is:** Larry J. Young et al., "Increased Affiliative Response to Vasopressin in Mice Expressing the Vasopressin Receptor from a Monogamous Vole," *Nature* 400 (1999): 766–768.

24 **knock-out mice:** Jennifer N. Ferguson et al., "Social Amnesia in Mice Lacking the Oxytocin Gene," *Nature Genetics* 25 (2000): 284–288.

25 **a kind of adaptive "tuning knob":** Larry J. Young and Elizabeth A.D. Hammock, "On Switches and Knobs, Microsatellites and Monogamy," *Trends in Genetics* 23 (2007): 209–212.

26 **nursing mothers awash in oxytocin:** Kerstin Uvnas-Moberg and Maria Eriksson, "Breastfeeding: Physiological, Endocrine, and Behavioral Adaptations Caused by Oxytocin and Local Neurogenic Activity in the Nipple and Mammary Gland," *Acta Paediatrica Scandinavica* 85 (1996): 525–530.

27 **New mothers are under a great deal of stress:** Kerstin Uvnas-Moberg, "Antistress Pattern Induced by Oxytocin," *News in Physiological Sciences* 13 (February 1998): 22–26.

28 **not all nursing mothers are alike:** Kerstin Uvnas-Moberg et al., "Personality Traits in Women Four Days Postpartum and Their Correlation with Plasma Levels of Oxytocin and Prolactin," *Journal of Psychosomatic Obstetrics and Gynecology* 2 (1990): 261–273.

28 **supported Uvnas-Moberg's findings:** Kathleen Light et al., "Oxytocin Responsivity in Mothers of Infants: A Preliminary Study of Relationships with Blood Pressure During Laboratory Stress and Normal Ambulatory Activity," *Health Psychology* 19 (2000): 560–567.

29 **primary maternal preoccupations:** James F. Leckman, Linda C. Mayes, and Donald J. Cohen, "Primary Maternal Preoccupation

Revisited: Circuits, Genes, and the Crucial Role of Early Life Experience," *Sabinet Online* 15 (2002): 4–12.

29 **intense visual contact:** G. C. Gonzaga et al., "Romantic Love and Sexual Desire in Close Relationships," *Emotion* 6 (2006): 163–179. Also see R. A. Turner et al., "Preliminary Research on Plasma Oxytocin in Healthy, Normal Cycling Women Investigating Emotion and Interpersonal Distress," *Psychiatry* 62 (1999): 97–113.

30 **a cow sees:** Kerstin Uvnas-Moberg, *The Oxytocin Factor* (New York: Da Capo, 2003), p. 94. Discussion of dairy cows.

30 **a mother's head:** Andreas Bartels and Semir Zeki, "The Neural Correlates of Maternal and Romantic Love," *NeuroImage* 21 (2004): 1155–1166.

30 **the image of her child:** Gareth Leng, Celine Caquineau, and Nancy Sabatier, "Regulation of Oxytocin Secretion," *Vitamins and Hormones* 71 (2005): 27–58.

30 **mere exposure:** Israel Liberzon et al., "Motivational Properties of Oxytocin in Conditioned Place Preference Paradigm," *Neuropsychopharmacology* 17 (1997): 353–359.

THREE — A Mind on Nature

34 **rare large mammal:** John Noble Wilford, "When Humans Became Human," *New York Times*, February 26, 2002. Review of Richard Klein with Blake Edgar, *The Dawn of Human Culture* (New York: Wiley, 2002).

34 **ancient artists:** Craig Packer and Jean Clottes, "When Lions Ruled France," *Natural History*, November 2000, p. 56.

35 **Jaynes made the case:** Julian Jaynes, *The Origin of Consciousness in the Breakdown of the Bicameral Mind* (Boston: Houghton Mifflin, 1976), pp. 29, 47, 453.

36 **Nadia's mysterious gift:** Nicholas Humphrey, "Cave Art, Autism, and the Evolution of the Human Mind," *Journal of Consciousness Studies* 6 (1999): 117–118. Also see David Pariser, "Nadia's Drawings: Theorizing About an Autistic Child's Phenomenal Ability," *Studies in Art Education* 22 (1981): 20–31.

37 **extended selective attention:** Lynn Waterhouse, Deborah Fein, and Charlotte Modahl, "Neurofunctional Mechanisms in Autism," *Psychological Review* 103 (1996): 457–489.

37 **Calendar calculating savants:** www.holidays-info.com/calendar/calendar_2190.html. Also see Herman H. Spitz, "Calendar Calculating 'Idiot Savants' and the Smart Unconscious," *New Ideas in Psychology* 13 (1995): 167–182.

37 **who could also predict:** "Astronomical Structures in Ancient Egypt," *Washington Post*, April 6, 1998. Also see Peter Lancaster Brown, *Megaliths and Masterminds* (New York: Scribner, 1979), p. 231; Alexander Marshak, "Exploring the Mind of Ice Age Man," *National Geographic*, January 1975, p. 65.

38 **early human mind:** Stephen Mithen, "Explaining the Early Human Mind," *British Archeology*, June 1996, p. 6.

38 **a thimbleful:** Jaynes, *Origin of Consciousnes,* p. 28.

39 **asked to locate a target number:** Spitz, "Calendar Calculating," p. 173.

39 **thin slicing:** Malcolm Gladwell, *Blink* (New York: Little, Brown, 2005), p. 23.

41 **the Foré tribe:** Jared Diamond, "New Guineans and Their Natural World," in *Biophilia Hypothesis*, ed. S. R. Kellert and E. O. Wilson (Washington, DC: Island, 1993), pp. 251–271.

FOUR — The Rules of Engagement

44 **infantile feeding experience:** Jack D. Caldwell, "Central Oxytocin and Female Sexual Behavior," *Annals of the New York Academy of Sciences* 652 (1992): 166–179

45 **affiliate behavior:** Kerstin Uvnas-Moberg, "Physiological and Endocrine Effects of Social Contact," *Integrative Neurobiology of Affiliation* 887 (1997): 146–163; Uvnas-Moberg, "Endocrinological Control of Food Intake," *Nutrition Reviews* 48 (1990): 57–63; Uvnas-Moberg, conversation with author, Washington, D.C., 1999.

46 **precursor versions of oxytocin:** James L. Goodson, "The Vertebrate Social Behavior Network: Evolutionary Themes and Variations," *Hormones and Behavior* 48 (2005): 11–22.

46 **microscopic roundworms:** Nicholas Wade, "Can Social Behavior of Man Be Glimpsed in a Lowly Worm?" *New York Times*, September 8, 1998.

46 **fish valiantly attempting to rescue:** E. W. Gudger, "Some Instances of Supposed Sympathy Among Fishes," *Scientific Monthly* 28 (1929): 266–271.

46 **nature cobbled together nine amino acids:** Maria Petersson, *Short- and Long-Term Cardiovascular and Behavioral Effects of Oxytocin* (Stockholm: Karolinska Institute, 1999), p. 19.

46 **sense a full belly:** Kerstin Uvnas-Moberg, *The Oxytocin Factor* (New York: Da Capo Press, 2003), pp. 23, 76, 146.

47 **less than twenty minutes in the bloodstream:** Cort Pedersen. Conversation with author, Washington , D. C., January 2007.

48 **paranoid tendencies of the amygdala:** Daniel Huber, Pierre Veinante, and Ron Stoop, "Vasopressin and Oxytocin Excite Distinct Neuronal Populations in the Central Amygdala," *Science* 308 (2005): 245–248. Also see K. M. Kramer et al., "Developmental Effects of Oxytocin on Neural Activation and Neuropeptide Release in Response to Social Stimuli," *Hormones and Behavior* 49 (2006): 206–214. J. N. Ferguson et al., "Oxtyocin in the Medial Amygdala Is Essential for Social Recognition in the Mouse," *Journal of Neuroscience* 20 (2001): 8278–8285.

48 **noradrenalin:** Maria Petersson et al., "Oxytocin Increases Locus Coeruleus Alpha 2-Adrenoreceptor Responsiveness in Rats," *Neuroscience Letters* 225 (1998): 115–118. M. Petersson et al., "Oxytocin Increases the Density of High Affinity A2-Adrenoceptors Within the Hypothalamus, the Amygdala, and the Nucleus of the Solitary Tract in Overiectomized Rats," *Brain Research*, July 12, 2005, pp. 234–239.

48 **HPA stress axis:** Richard Windle et al., "Oxytocin Attenuates Stress-Induced c-*fos* mRNA Expression in Specific Forebrain Regions Associated with Modulation of Hypothalmo-Pituitary-Adrenal Activity," *Journal of Neuroscience,* March 24, 2004, pp. 2974–2982.

49 **brain networks:** Cort A. Pedersen and Maria L. Boccia, "Oxytocin Links Mothering Received, Mothering Bestowed, and Adult Stress Responses," *Stress* 5 (2002): 259–267.

49 **instigator of fight/flight:** C. Sue Carter and Margaret Altemus, "Integrative Functions of Lactational Hormones in Social Behavior and Stress Management," *Integrative Neurobiology of Affiliation* 807 (1997): 164–174. Also see Geert J. Devries and Constaza Villalba, "Brain Sexual Dimorphism and Sex Differences in Parental and Other Social Behaviors," *Integrative Neurobiology of Affiliation* 807 (1997): 273–282.

49 **lowers their HPA chemistry:** A. Stachowiak et al., "Effects of Oxytocin on the Function and Morphology of the Rat Adrenal

Cortex: In Vitro and In Vivo Investigations," *Research in Experimental Medicine* 195 (1995): 265–274. Also see Gerald Gimpl and Falk Fahrenholz, "The Oxytocin Receptor System: Structure, Function, and Regulation," *Physiological Reviews* 81 (2001): 629–683.

49 **rats to venture into:** K. Uvnas-Moberg et al., "Oxytocin Reduces Exploratory Motor Behavior and Shifts Activity Towards the Center of the Arena in Male Rats," *Acta Physiologica Scandinavica* 145 (1992): 429–430. Also see M. Petersson, M. Eklund, and K. Uvnas-Moberg, "Oxytocin Decreases Corticosterone and Nociception and Increases Motor Activity in OVX Rats," *Maturitas* 51 (2005): 426–433.

49 **dopamine-producing brain cells:** M. D. Smeltzer et al., "Dopamine, Oxytocin, and Vasopressin Receptor Binding in the Medial Prefrontal Cortex of Monogamous and Promiscuous Voles," *Neuroscience Letters* 394 (2006): 146–151.

50 **positive interactions:** V. Hillegaart, P. Alster, and K. Uvnas-Moberg, "Sexual Motivation Promotes Oxytocin Secretion in Male Rats," *Peptides* 19 (1998): 39–45.

50 **interactions involve welcome touch:** K. C. Light, K. M. Grewen, and J. A. Amico, "More Frequent Partner Hugs and Higher Oxytocin Levels Are Linked to Lower Blood Pressure and Heart Rate in Premenopausal Women," *Biological Psychology* 69 (2005): 5–21. Also see Uvnas-Moberg, *Oxytocin Factor*, pp. 107–116.

50 **Oxytocin and Opioids:** M. Zubryzcka et al., "Inhibition of Trigemino-Hypoglossal Reflex in Rats by Oxytocin Is Mediated by U And K Opioid Receptors," *Brain Research* 1035 (2005): 67–72.

50 **feel-good chemicals, dopamine, serotonin, and beta endorphin:** M. Galfi et al., "Serotonin-Induced Enhancement of Vasopressin and Oxytocin Secretion in Rat Neurohypohyseal Tissue Culture," *Regulatory Peptides* 127 (2005): 225–231.

50 **state of calm connect:** "Oxtyocin May Mediate the Benefits of Positive Social Interaction and Emotions," *Psychoneuroendocrinology* 23 (1998): 819–835.

FIVE — Brave New World

52 **hunting camps:** Donald Johnson and Blake Edgar, *From Lucy to Language* (New York: Simon & Schuster, 1996), p. 99.

53 **behavior radically shifted:** Jared Diamond, "The Great Leap Forward," *Discover,* May 1989, pp. 50–60. Also see John Pfeiffer, "Man the Hunter," *Horizon* 23 (1971): 29–33.

54 **pelvis that supported upright posture:** *The Human Dawn: Timeframe* (Alexandria, VA: Time-Life Books, 1990), pp. 31–32.

54 **Meredith Small explains:** Meredith F. Small, "Our Babies, Ourselves," *Natural History,* October 1997, p. 48.

55 **social desirability:** B. Sjogren et al., "Changes in Personality Pattern During the First Pregnancy and Lactation," *Journal of Psychosomatic Obstetric Gynocology* 21 (2000): 31–38.

56 **first midwives:** Wendy R. Trevathan, *Human Birth: An Evolutionary Perspective* (New York: Aldine de Gruyter, 1987). Also see Marshall Klaus, "Touching During and After Childbirth," in *Touch in Early Development*, ed. Tiffany M. Field (Mahwah, NJ: Lawrence Erlbaum 1995), pp. 19–33.

56 **apes are attentive to the sights and sounds of newborns:** "Evolutionary Context of Human Development: The Cooperative Breeding Model," in *Attachment and Bonding: A New Synthesis*, ed. C. S. Carter et al. (Cambridge: MIT Press, 2006), pp. 9–32.

56 **distress calls:** Katsuhiko Nishimori et al., "Pervasive Social Deficits, but Normal Parturition, in Oxytocin Receptor-Deficient Mice," *PNAS* 102 (2005): 16096–16101.

57 **pup licking:** D. Lui et al., "Maternal Care, Hippocampal Glucocotocoid Receptors, and Hypothalamic-Pituitary-Adrenal Response to Stress," *Science* 227 (1997): 1659–1662. Also see Caldji et al., "Maternal Care During Infancy Regulates the Development of Neural Systems Mediating the Expression of Fearfulness in the Rat," *Proceeds of the National Academy of Sciences, USA* 95 (April 1998): 5335–5340.

57 **went on to be more maternal:** Alison Flemming, "Mothering Begets Mothering in Rat and Human Mothers," Frontiers in the Study of Individual Variation: Impact of Relationships, Fourth Interdisciplinary Dialogue, Bowen Center for Family Study, April 22–23, 2006.

58 **discussion of maternal behavior:** D. D. Francis, F. C. Champagne, and M. J. Meaney, "Variations in Maternal Behavior Are Associated with Differences in Oxytocin Receptor Levels in the Rat," *Journal of Neuroendocrinology* 12 (2000): 1145–1148. Also

see C. A. Pedersen et al., "Maternal Behavior Deficits in Nulliparous Oxytocin Knockout Mice," *Genes, Brain, and Behavior* 5 (2006): 274–281; C. A. Pedersen and M. L. Boccia, "Oxytocin Links Mothering Received, Mothering Bestowed, and Adult Stress Responses," *Stress* 5 (2002): 259–267.

58 **oxytocin tilts the balance of oral grooming:** Cort A. Pedersen, "How Love Evolved from Sex and Gave Birth to Intelligence and Human Nature," *Journal of Bioeconomics* 6 (January 2004): 39–63.

59 **growing sense of trust:** Paul J. Zak, Robert Kurzban, and William T. Matzner, "Oxytocin Is Associated with Human Trustworthiness," *Hormones and Behavior* 48 (2005): 522–527.

60 **study conducted by Swiss neuroeconomists:** Michael Kosfeld et al., "Oxytocin Increases Trust in Humans," *Nature* 435 (2005): 673–676.

61 **shown to permeate the brain:** Jan Born et al., "Sniffing Neuropeptides: A Transnasal Approach to the Human Brain," *Nature Neuroscience* 5 (2002): 514–516.

62 **stable, supportive relationships:** Karen M. Grewen et al., "Effects of Partner Support on Resting Oxytocin, Cortisol, Norepinephrine, and Blood Pressure Before and After Warm Partner Contact," *Psychosomatic Medicine* 67 (2005): 531–538.

62 **enhance the production of oxytocin:** A. E. Storey et al., "Hormonal Correlates of Paternal Responsiveness in New And Expectant Fathers," *Evolution of Human Behavior* 21 (2000): 79–95.

62 **attentive fathers:** Jeffrey A. French, "Variation in Parental Care in a Monogamous Primate: Causes and Long-Term Consequences," Frontiers in the Study of Individual Variation: Impact of Relationships, Fourth Interdisciplinary Dialogue, Washington, D.C., April 22–23, 2006. Also see Janet K. Bester-Meredith and Catherine A. Marler, "Vasopressin and the Transmission of Paternal Behavior Across Generations in Mated, Cross-Fostered Peromyscus Mice," *Behavioral Neuroscience* 117 (2003): 455–463.

62 **capacity for altruism:** Ariel Knafo et al., "Individual Differences in Allocation of Funds in the Dictator Game Associated with Length of the Arginine Vasopressin 1a Receptor (AVPR1a) RS3 Promoter-Region and Correlation Between RS3 Length and Hippocampal mRNA," *Genes, Brains, and Behavior,* Online Early Articles, Doi:10.111/j.1601–183X2007.00341.x

63 **vasopressin receptor and human pair-bonding:** Hasse Walum et al., "Genetic Variation in the Vasopressin Receptor 1a Gene (AVPR1A) Associates with Pair-Bonding Behavior in Humans," *PNAS* 105 (September 2008): 14153–14156.

63 **Oxytocin instead fosters courage:** *The Oyxtocin Factor* (New York: Da Capo, 2003), p. 66.

SIX — Good Dog

68 **wolf skulls facing out:** Juliet Clutton-Brock, *Domesticated Animals from Early Times* (Austin: University of Texas Press, 1981), p. 11.

68 **species adapted for such a niche:** Jerison, *Brain Size*, p. 83; Yi-Fu Tuan, *Dominance and Affection* (New Haven, CT: Yale University Press, 1984), p. 77.

68 **400,000 years ago:** Nicholas Wade, "From Wolf to Dog, Yes, But When?" *New York Times*, November 22, 2002.

68 **wolves driving a herd of caribou:** Frederick E. Zeuner, *A History of Domesticated Animals* (New York: Harper & Row, 1963), p. 62.

69 **Wolves are watchful:** Benson E. Ginsburg and Laurie Hiestand, "Humanity's 'Best Friend': The Origins of Our Inevitable Bond with Dogs," in *The Inevitable Bond*, ed. Hank Davis and Dianne Balfour (Cambridge: Cambridge University Press, 1992), pp. 93–108.

69 **Badgers and coyotes:** Steven C. Minta, Kathryn A. Minta, and Dale Lott, "Hunting Associations Between Badgers and Coyotes," *Journal of Mammology* 73 (1992): 814–820.

70 **interested in marrow:** Sally Fallon and Mary G. Enig, "Caveman Cuisine," Weston A. Price Foundation, www.westonaprice.org/traditional_diets/caveman_cuisine.html.

70 **lured wolves into cave dwellings:** C. Vila, et al., Multiple and Ancient Origins of the Domestic Dog, *Science* 276 (1997): 1687–1689.

74 **stroking a rat:** Irene Lund et al., "Sensory Stimulation (Massage) Reduced Blood Pressure in Unanesthetized Rats," *Journal of the Autonomic Nervous System* 78 (1999): 30–37. Also see Greta Agren et al., "The Oxytocin Antagonist 1-deamino-2-D-Try-(Oet)-4-Thr-8-Orn-oxytocin Reverses the Increase in Withdrawal Response Latency to Thermal But Not Mechanical Nociceptive Stimuli Following Oxytocin Administration or Massage-Like

Stroking in Rats," *Neuroscience Letters* 187 (1995): 49–52; Kerstin Uvnas-Moberg, "Physiological and Endocrine Effects of Social Contact," in *The Integrative Neurobiology of Affiliation* (New York: New York Academy of Sciences 1997), pp. 146–163.

74 **owners' blood levels:** J. S. J. Odendaal and R. A. Meintjes, "Neurophysiological Correlates of Affiliative Behavior Between Humans and Dogs," *Veterinary Journal* 165 (2003): 296–301.

74 **suckling a wolf cub:** James A. Serpell, "Pet Keeping in Non-Western Societies: Some Popular Misconceptions," in *Animals and People Sharing the World*, ed. Andrew N. Rowan (Hanover, NH: University Press of New England), pp. 33–52. Serpell also cites W. E. Roth, "An Introductory Study of the Arts, Crafts and Customs of the Guiana Indians," *Annual Report Bureau of American Ethnology* 38 (1934): 551–556.

75 **bundle of jangled nerves:** Keeler et al. (1990), as quoted in Grandin and Deesing, "Behavioral Genetics," p. 3.

76 **Even the first generation:** Zeuner, *History of Domesticated Animals,* p. 104.

77 **Belyaev's friendly foxes:** Lyudmila N. Trut, "Early Canid Domestication: The Farm-Fox Experiment," *American Scientist* 87 (1999): 160–169. Also see Temple Grandin and Mark J. Deesing, "Behavioral Genetics and Animal Science," in *Genetics and the Behavior or Domestic Animals,* ed. Temple Grandin (San Diego: Academic, 1998), pp. 1–30.

77 **cort levels have dropped:** A. Stachowiak et al., "Effects of Oxytocin on the Function and Morphology of Rat Adrenal Cortex: In Vitro and In Vivo Investigations," *Research in Experimental Medicine* 195 (1995): 265–274.

77 **more affectionate people:** Malcolm W. Browne, "New Breed of Fox As Tame As a Pussycat," *New York Times,* March 30, 1999.

78 **gene called RUNX2:** John W. Fodon III and Harold Garner, "Detection of Length-Dependent Effects on Tandem Repeat Alleles by 3-D Geometric Decomposition of Craniofacial Variation," *Development, Genes, and Evolution* 217 (2007): 79–85. Also see Yechezkel Kashi and David G. King, "Simple Sequence Repeats as Advantageous Mutators in Evolution," *Trends in Genetics* 22 (2006): 253–259.

79 **feeding wild animals:** Dale F. Lott, "Feeding Wild Animals: The Urge, the Interaction, and the Consequences," *Anthrozoos* 1 (1988): 255–257.

79 **Primal ego boost:** Brian Hayden, "A New Overview of Domestication," in *Last Hunters, First Farmers*, ed. D. T. Price and A. B. Gebauer (Santa Fe, NM: School of American Research Press, 1995), p. 296.

79 **The little dogges:** Shakespeare, *King Lear* 3.6.

SEVEN — On the Shoulders of Giants

82 **ancient mental talent:** Stephen Mithen, "Cave Art, Autism, and the Evolution of the Human Mind," *Journal of Consciousness Studies* 6 (1999): 128–131.

82 **learning rules:** Edward O. Wilson, "Biophilia and the Conservation Ethic," in *The Biophilia Hypothesis,* ed. Stephen R. Kellert and Edward O. Wilson (Washington, DC: Island, 1993), p. 32.

83 **horse disappeared in North America:** Juliet Clutton-Brock, *Domesticated Animals from Early Times* (Austin: University of Texas Press, 1981), p. 81.

83 **tamed a breed of horse:** Sandra L. Olsen, "Hoofprints," *Natural History,* May 2008, pp. 26–32.

84 **their only blind spot:** Steven Budiansky, *The Nature of Horses* (New York: Free Press, 1997), p. 111.

85 **Ray Hunt:** Personal conversations with author, 1993–1994. Also see Hunt, *Think Harmony with Horses* (Fresno, CA: Pioneer, 1985). Hunt, *Think Harmony*, p. 13. Also see Barbara Hague, *Turning Loose* (McNabb & Connelly, 1994). Video recording.

85 **Horses for all their might:** Budiansky, *Nature of Horses*, p. 93.

86 **gentle method:** E. C. Marchant, ed., *Xenophon* (London: Heinemann, 1925); anger, p. 325; gentle bit, p. 343; dates, p. 184.

88 **Clever Hans:** Oskar Pfungst, *Clever Hans: The Horse of Mr. Von Osten* (New York: Holt, Rinehart & Winston, 1965); befriend, p. 31; head movement, pp. 47–49, 104; thou shalt, p. 91.

91 **mini round-pen experience:** C. S. Carter, "Biological Perspectives on Social Attachment and Bonding," in *Attachment and Bonding: A New Synthesis,* ed. C. S. Carter et al. (Cambridge: MIT Press, 2006), p. 94. Also see A. Courtney DeVries et al., "The Effects of Stress on Social Preferences Are Sexually Dimorphic in Prairie Voles," *Proceedings of the National Academy of Sciences of the United States of America* 93 (1996): 11980–11984.

92 **nonnoxious, soothing contact:** Irene Lund et al., "Sensory Stimulation (Massage) Reduces Blood Pressure in Unanesthetized

Rats," *Journal of the Autonomic Nervous System* 78 (1999): 30–37. Also see Uvnas-Moberg et al., "Stroking of the Abdomen Causes Decreased Locomotor Activity in Conscious Male Rats," *Physiology and Behavioral Science* 60 (1995): 1–3.

94 **herd indulged themselves in a bout of grooming:** Claudia Feh and Jeanne de Mazieres, "Grooming at a Preferred Site Reduces Heart Rate in Horses," *Journal of Animal Behavior* 46 (1993): 1191–1194. Also see Claudia Feh, "Relationships and Communication in Socially Natural Horse Herds" (unpublished paper).

95 **unknown class of sensory nerves:** Kerstin Uvnas-Moberg, "Physiological and Endocrine Effects of Social Contact," in *The Integrative Neurobiology of Affiliation* (New York: New York Academy of Sciences, 1997), pp. 146–163.

96 **first equestrians:** J. K. Anderson, *Ancient Greek Horsemanship* (Berkley: University of California Press, 1961).

98 **Seatsbones:** Anderson, *Ancient Greek Horsemanship*, p. 62.

99 **ideomotor actions:** Evon Z. Vogt and Ray Hyman, *Water Witching USA* (Chicago: University of Chicago Press, 1959), p. 132. Also see Vogt and Hyman, *Water Witching,* pp. 135–136; William James, *The Principles of Psychology* (Chicago: William Benton, 1952), pp. 792–793.

100 **muscle reading:** Vogt and Hyman, *Water Witching,* p. 103.

101 **free won't:** Benjamin Libet, *Neurophysiology of Consciousness* (Boston: Birkhauser, 1993). Also see Elsa Youngstead, "Case Closed for Free Will?" *Science NOW Daily News*, April 14, 2008.

102 **minute responses initiated by thoughts:** Vogt and Hyman, *Water Witching,* p. 140.

102 **animating artificial limbs:** Luke Arm, www.60secondscience.com/archive/health-news-articles-medicine-news/dean-kamens-luke-arm-bionic-pr.php.

103 **gallop forward into the noise and confusion:** Elizabeth A. Lawrence, "Horses in Society," in *Animals and People Sharing the World,* ed. Andrew N. Rowan (Hanover, NH: University Press of New England, 1988), p. 105.

EIGHT — The Meeting of the Minds

106 **Wolf bones:** Juliet Clutton-Brock, "Origins of the Dog: Domestication and Early History," in *The Domestic Dog*, ed. James Serpell (Cambridge: Cambridge University Press, 1995), pp. 7–20.

107 **vocal convergence:** Eugene S. Morton, "On the Occurrence and Significance of Motivation-Structural Rules in Some Bird and Mammal Sound," *American Naturalist* 3 (1997): 855–869.

107 **paralanguage:** Peter Farb, *Wordplay* (New York: Knopf, 1973), p. 62.

107 **Articulate language:** Charles Darwin, *The Descent of Man: And Selection in Relation to Sex* (London: John Murray, 1882), p. 85.

108 **had never been sued:** Nalini Ambady et al., "Surgeons' Tone of Voice: A Clue to Malpractice History," *Surgery* (July 2002): pp. 5–9.

108 **human talking makes sense:** Mary Midgely, *Beast and Man* (Ithaca, NY: Cornell University Press, 1978), p. 234.

110 **read . . . the owner's face:** Paul Ekman, *Emotions Revealed* (New York: Holt, 2003).

111 **oxytocin could help humans "read minds":** Gergor Domes et al., "Oxytocin Improves 'Mind-Reading' in Humans," *Biological Psychiatry* 61 (2007): 731–733.

111 **women are better than men at deciphering:** Matthew D. Lieberman, "Intuition: A Social Cognitive Neuroscience Approach," *Psychological Bulletin* 126 (2000): 109–137.

112 **intravenous doses of oxytocin:** "Autism Research Efforts Highlighted in Biological Psychiatry Special Issue," *Science Update*, NIMH, February 6, 2007.

112 **paralinguistic communication:** Gregory Bateson, *Steps to an Ecology of Mind* (New York: Ballantine, 1972), p. 371.

113 **vocal equipment:** Steven Pinker, *The Language Instinct* (New York: Morrow, 1994), p. 265.

113 **Human grunts:** Lorraine McCune et al., "Grunt Communication in Human Infants," *Journal of Comparative Psychology* 110 (1996): 27–37.

113 **as vervets mature:** Dorothy L. Cheney and Robert M. Seyfarth, *How Monkeys See the World* (Chicago: University of Chicago Press, 1990), pp. 115–120.

114 **vocal link:** U. Jurgens and K. Hammerschmidt, "Common Acoustic Features in the Vocal Expressions of Emotions in

Monkeys and Man" (presentation to the Fifteenth Annual Meeting of the IBNS, May 2006).

114 **Barbara Smuts:** Barbara Smuts, "Encounters with Animal Minds," *Journal of Consciousness Studies* 8 (2001): 293–309.

NINE — The Dog of the Hare

122 **unique ability to understand:** Brian Hare et al., "The Domestication of Social Cognition in Dogs," *Science* 298 (2002): 1634–1636. Also see Brian Hare and Michael Tomasello, "Human-like Social Skills in Dogs?" *Trends in Cognitive Sciences* 9 (2005): 439–444; Kate Douglas, "Mind of a Dog," *New Scientist*, March 4, 2000, pp. 20–27.

123 **just as good as dogs:** Brian Hare et al., "Social Cognitive Evolution in Captive Foxes Is a Correlated By-Product of Experimental Domestication," *Current Biology* 15 (2005): 226–230.

123 **fear and aggression:** Nicholas Wade, "Nice Rats, Nasty Rats: Maybe It's All in the Genes," *New York Times*, July 25, 2006.

123 **at least sixty more words:** Stanley Coren, *The Intelligence of Dogs* (New York: Bantam, 1994), p. 97.

123 **motherese:** Marc D. Hauser, *The Evolution of Communication* (Cambridge: MIT Press, 1996), pp. 332–334.

124 **acoustical profile:** Patricia B. McConnell, "Acoustic Structure and Receiver Response in Domestic Dogs, *Canis familiaris*," *Association for the Study of Animal Behavior* 39 (1990): 897–904.

124 **Rico now understands the meaning of two hundred words:** Juliane Kaminski, Josep Call, and Julia Fischer, "Word Learning in a Domestic Dog: Evidence for 'Fast Mapping,'" *Science* 304 (2004): 1682–1683.

126 **communicative and social abilities:** Paul Bloom, "Can a Dog Learn a Word?" *Science* 304 (2004), 1605–1606.

126 **dog owners responded to their dog's soliciting stares:** Miho Nagasawa et al., "Dog's Gaze to His Owner Functions as Social Attachment and Increases Owner's Urine Oxytocin" (poster presentation at the Canine Science Forum 2008, Budapest, Hungary).

126 **Linguist Mark Feinstein:** Stephen Budiansky, "What Animals Say to Each Other," *U.S. News and World Report*, June 5, 1995.

127 **feel safer and sleep better:** Benedict Carey, "An Active Purposeful Machine That Comes Out to Play at Night," *New York Times*,

October 23, 2007; Colin Barras, "How Nodding Off Can Help You Remember," *New Scientist*, February 23, 2008, p. 13.

128 **long, slow alpha-wave brain action:** Kerstin Uvnas-Moberg et al., "High Doses of Oxytocin Cause Sedation and Low Doses Cause an Anxiolytic-like Effect in Male Rats," *Pharmacology of Biochemistry and Behavior* 49 (1994): 101–106. Also see Liesbeth van Londen et al., "Plasma Arginine Vasopressin and Motor Activity in Major Depression," *Society of Biological Psychiatry* 43 (1998): 196–204.

128 **animals are good to think:** Claude Levi-Strauss, *Totemism* (Boston: Beacon, 1962), p. 89.

128 **distinguish the emotional state:** Peter Pongracz, Csaba Molnar, and Adam Miklosi, "Acoustic Parameters of Dog Barks Carry Emotional Information for Humans," *Applied Animal Behavior Science* 100 (2006): 228–240.

128 **barks are meaningful noise:** Sophia Yin, "A New Perspective on Barking in Dogs," *Journal of Comparative Psychology* 116 (2002): 189–193. Also see Sophia Yin and Brenda McCowan, "Barking in Domestic Dogs: Context Specificity and Individual Identification," *Animal Behavior* 68 (2004): 343–355.

128 **wag a tail:** Sandra Blakeslee, "If You Want to Know If Spot Loves You So, It's in His Tail," *New York Times*, April 24, 2007.

129 **play-bow:** Marc Bekoff and Collin Allen, "Intentional Communication and Social Play: How and Why Animals Negotiate and Agree to Play," in *Animal Play: Evolutionary, Comparative, and Ecological Perspectives*, ed. Marc Bekoff and John A. Byers (Cambridge: Cambridge University Press, 1998), pp. 97–114.

129 **Non-verbal communication:** Juliet Clutton-Brock, *Domesticated Animals from Early Times* (Austin: University of Texas Press, 1981), p. 9.

129 **making oneself understood:** Mary Midley, *Beast and Man* (Ithaca, NY: Corness University Press, 1978), p. 234.

131 **some 400 million strong:** Dierdre van Dyk, "The Mother of All Dogs," *Time*, December 2, 2002, pp. 78–79.

131 **the Hare clan:** Joel S. Savishinsky, *The Trail of the Hare: Environment and Stress in a Sub-Arctic Community* (New York: Gordon & Breach, 1994); "Processing the Environment," p. 58; good with dogs, p. 209; diminished nomads, p. 192.

133 **the dog's superior sense of smell:** Patricia B. McConnell, *The Other End of the Leash* (New York: Ballantine, 2002), p. 72.

136 **his hand gently caressing a puppy:** Karen E. Lange, "Dogs, A Love Story," *National Geographic Magazine*, January 2002, p. 4.

TEN — New Game

138 **happened first in the Levant:** Ofer Bar-Yosef, "The Natufian Culture in the Levant; Threshold to the Origins of Agriculture," *Evolutionary Anthropology* 6 (1998): 159–174.

138 **rudimentary farming:** Colin Tudge, *Neanderthals, Bandits, and Farmers* (New Haven: Yale University Press, 1999).

141 **the Hidasta:** Gilbert L. Wilson, *Buffalo Bird Woman's Garden* (St. Paul: Minnesota Historical Society Press, 1987); tools, p. 13; singing to the corn, p. 27.

141 **plants . . . capable of social recognition:** Susan A. Dudley and Amanda File, "Kin Recognition in an Annual Plant," *Biology Letters,* June 13, 2007, http://publishing.royalsociety.org/index.cfm?page=1566.

141 **[plants] respond to human touch:** James F. Cahill Jr., Jeffrey P. Castelli, and Brenda B. Casper, "Separate Effects of Human Visitation and Touch on Plant Growth and Herbivory in an Old-Field Community," *American Journal of Botany* 89 (2002): 1401–1409.

141 **skeletons of the first farmers:** John Noble Wilford, "In Ancient Bones Researchers See Pictures of Health," *New York Times*, September 7, 1999.

142 **alive but not thriving:** Juliet Clutton-Brock, *Domesticated Animals from Early Times* (Austin: University of Texas Press, 1981), p. 68.

142 **memories:** Peter Rowley-Conwy, "In Sorrow Thou Shalt Eat All Thy Days," *British Archeology,* February 1997, p. 7. Also see Jared Diamond, *The Third Chimpanzee* (New York: HarperCollins, 1992), pp. 185–187.

142 **small kinship bonds:** John Pfeiffer, "How Man Invented Cities," *Horizon* 4 (Autumn 1972): 13–18. Also see, Diamond, *Third Chimpanzee,* p. 237.

142 **surviving aboriginal societies:** Merlin Donald, *Origins of the Modern Mind* (Cambridge, MA: Harvard University Press, 1991), p. 139.

143 **earliest cities:** Lewis Mumford, "The City in History," *Horizon* 3 (July 1961): 41.

143 **Catalhoyuk:** James Mellaart, "Man's First Murals," *Horizon* 5 (September 1962): 11–13; Ian Hodder, *The Leopard's Tale* (London: Thames & Hudson, 2006); population, p. 95; housing density, p. 100; murals and animal decoration, pp. 47, 197–199; public/private identity, pp. 92–95; burials, p. 24; domestic versus wild animals, pp. 84, 197–200; dogs, p. 84; leopards, pp. 9–12; bones, p. 11; shift to domestication, pp. 249–251; cattle bones, p. 255.

144 **aurochs:** Jared Diamond, "Why Is a Cow Like a Pyramid?" *Natural History,* July 1995, p. 12.

147 **showing no mercy:** J. M. C. Toynbee, *Animals in Roman Life and Art* (New York: Cornell University Press, 1968), p. 148.

148 **two-legged salt-lick:** Frederick Zeuner, *A History of Domesticated Animals* (New York: Harper & Row, 1963); ungulates love salt, p. 119; Valerius Geist quote, p. 291.

149 **women writ large:** Lewis Mumford, "The City in History," *Horizon* 3 (1961): 37.

149 **woman sitting between two leopards:** Hodder, *Leopard's Tale,* pp. 208–209.

ELEVEN — Made for Each Other

152 **newly domesticated humans:** Brenda J. Baker, "Secrets in the Skeletons," *Archeology,* May–June, 2001, p. 47.

152 **diseases and parasites:** Charlotte Roberts, "Never a Society That Suffered No Illness," *British Archeology,* December 1995, pp. 8–9.

152 **new social stress:** Frans B. M. de Waal, Filippo Aureli, and Peter G. Judge, "Coping with Crowding," *Scientific American*, May 2000, pp. 76–81.

152 **reputation and social worth:** "The Human Brain Appears Hard-Wired for Hierarchy," *NIMH Science News*, April 23, 2008; Nikhil Swaminathan, "For the Brain Cash Is Good, Status Is Better," *Scientific American News*, April 24, 2008; Elizabeth Lawrence, *Hoof Beats and Society* (Bloomington: Indiana University Press, 1985), p. 5.

153 **Growing beautiful and delicious things:** Charles A. Lewis obituary, *New York Times*, January 11, 2004.

153 **sense of control:** Lawrence, *Hoofbeats and Society*, p. 5.

154 **tolerance for pain:** Kerstin Uvnas-Moberg et al., "Anti-Nociceptive Effect of Non-Noxious Sensory Stimulation Is Partly Mediated Through Oxytocinergic Mechanisms," *Acta Physiologica Scandinavica* 149 (1993): 199–204. Also see Y. Miranda-Cardenas et al., "Oxytocin and Electrical Stimulation of the Paraventricular Hypothalamic Nucleus Produce Antinociceptive Effects That Are Reversed by an Oxytocin Antagonist," *Pain* 122 (May 2006): 182–189.

154 **heal their wounds:** Maria Petersson et al., "Oxytocin Increases the Survival of Musculocutaneous Flaps," *Arch Pharmacol* 357 (1998): 701–704.

154 **can even prevent:** Sevgin Ozlem Iseri et al., "Oxytocin Protects Against Sepsis-Induced Multiple Organ Damage: The Role of Neutrophils," *Journal of Surgical Research* 126 (June 2005): 73–81.

155 **cow must be quite relaxed:** Juliet Clutton-Brock, *Domesticated Animals from Early Times* (Austin: University of Texas Press, 1981), p. 67.

156 **cow forming an attachment to humans:** Paul Hemsworth, John Barnett, and Graham Coleman, "Fear of Humans and Its Consequences for the Domestic Pig," in *The Inevitable Bond*, ed. Hank Davis and Dianne Balfour (Cambridge: Cambridge University Press, 1992), pp. 274–275.

156 **convinced cows they were worthy of their milk:** Frederick J. Simoons, "The Antiquity of Dairying in Asia and Africa," *Geographical Review* 61 (1971): 431–439.

156 **tomb art:** Clutton-Brock, *Domesticated Animals,* p. 9; Frederick E. Zeuner, A *History of Domesticated Animals* (New York: Harper & Row, 1963), p. 224.

156 **tears:** Wolfgang Jagla et al., "Co-localization of TFF3 Peptide and Oxytocin in the Human Hypothalamus," *FASEB Journal* 14 (2000): 1126–1131.

156 **Tess:** Thomas Hardy, *Tess of the D'Urbervilles* (New York: Penguin, 1984), p. 177. As cited by Aaron H. Katcher and Alan M. Beck, "Health and Caring for Living Things," *Anthrozoos* 1 (Winter 1987): 175–183.

158 **Egyptians had developed:** Jared Diamond, "Zebras and the Anna Karenina Principle," *Natural History*, September 1994, p. 5. Also see Clutton-Brock, *Domesticated Animals*, p. 176; Heinrich Heideger, *Wild Animals in Captivity* (New York: Dover, 1965), p. 165.

158 **lion:** Anahad O'Connor, "Discovery Shows Sacred Status of Egyptian Lion," *New York Times,* January 20, 2004.

158 **No Native American:** Jared Diamond, *The Third Chimpanzee* (New York: HarperCollins, 1992), p. 240.

159 **full-blown domestication:** Jared Diamond, "Zebras and the Anna Karenina Principle," *Natural History*, September 1994, pp. 4–10.

159 **assimilation tendency:** Hediger, *Wild Animals,* p. 163.

159 **mistaken identity:** Fritz R. Walther, *Communication and Expression in Hoofed Mammals* (Bloomington: Indiana University Press, 1984), pp. 380–382.

160 **cats**: John Pickrell, "Oldest Known Pet Cat? 9,500-Year-Old Burial Found on Cyprus," *National Geographic*, April 8, 2004.

160 **cat ownership:** Eileen B. Karsh and Dennis Turner, "The Human-Cat Relationship." *In The Domestic Cat*, edited by Dennis Turner and Patrick Bateson (Cambridge: Cambridge University Press, 2000), p. 172.

161 **The cat goddess Bastet:** James A. Serpell, "The Domestication and History of the Cat," in *The Domestic Cat*, ed. Dennis Turner and Patrick Bateson (Cambridge: Cambridge University Press, 2000), p. 154.

161 **mummified cats:** Muriel Beadle, *The Cat: History, Biology, Behavior* (New York: Simon & Schuster, 1977), p. 69; Henry Fountain, "Even in Death, These Cats Got Special Treatment," *New York Times,* September, 27, 2004.

161 **The big human brain:** Personal conversations with Cort Pedersen, July, 2006.

162 **The Egyptians loved animals:** Kenneth Clark, *Animals and Men* (New York: William Morrow, 1977), pp. 15–16.

162 **biotechnological consequences:** Colin McGinn, review of Francis Fukuyama, *Our Posthuman Future: Consequences of the Biotechnology Revolution* (New York: Farrar, Strauss & Giroux, 2002), *New York Times Book Review*, May 5, 2002, p. 11.

TWELVE — The Survivors

166 **Nubia:** David Roberts, "Out of Africa: The Superb Artwork of Ancient Nubia," *Smithsonian Magazine,* June 1993, pp. 90–100.

167 **Sacred bulls:** C. W. Towne and E. N. Wentworth, *Cattle and Men* (Norman: University of Oklahoma Press, 1955), pp. 43–49.

168 **the Nuer:** Raymond C. Kelly, *The Nuer Conquest: The Structure and Development of an Expansionist System* (Ann Arbor: University of Michigan Press, 1985), p. 11.

168 **Evans-Pritchard:** John Ryle, review of Sharon E. Hutchinson, *Nuer Dilemmas* (Berkeley: University of California Press, 1997), *TLS*, May 23, 1997.

169 **They love their cattle:** Yi-Fu Tuan, *Dominance and Affection* (New Haven, CT: Yale University Press, 1984), pp. 91–92.

170 **the Dinka:** Godfrey Lienhardt, *Divinity and Experience: The Religion of the Dinka* (London: Oxford University Press, 1961), p. 135.

170 **Both of these great:** Stephen Buckley, "Loss of Culturally Vital Cattle Leaves Dinka Tribe Adrift in Refugee Camps," *Washington Post*, August 24, 1997.

170 **cow dance:** Frederick E. Zeuner, *A History of Domesticated Animals* (New York: Harper & Row, 1963), pp. 226–227.

171 **the Fulani:** H. A. S. Johnston, *The Fulani Empire of Sokoto* (London: Oxford University Press, 1970).

171 **Fulani live in clans:** Dale F. Lott and Benjamin L. Hart, "Applied Ethology in a Nomadic Cattle Culture, *Applied Animal Ethology* 5 (1979): 309–319. Also see Dale F. Lott and Benjamin L. Hart, "The Fulani and Their Cattle: Applied Behavioral Technology in a Nomadic Cattle Culture and Its Psychological Consequences," *National Geographic Society Research Reports* 14 (1982): 425–430.

175 **behavioral properties of the cattle:** Lott and Hart, "Fulani and Their Cattle," p. 429.

176 **Raika:** Ilse Kohler-Rollefson, "Camels in the Lands of Kings," *Natural History,* March 1995, pp. 55–60. Ilsa Kohler Rollefson, personal correspondence with author, May 1997.

176 **They made a camel:** James C. Simmons, "The Arabian Desert Is No Place for Camels," *Audubon,* January 1991, pp. 38–45. Also see Edward Hymans, *Animals in the Service of Man* (London: Dent, 1972), p. 128; Hilde Gauthier-Pilters and Anne Innis Dagg, *The Camel: Its Evolution, Behavior, and Relationship to Man* (Chicago: University of Chicago Press, 1981); Julia Clutton-Brock, *Domesticated Animals from Early Times* (Austin: University of Texas Press, 1981), pp. 121–125.

178 **animals who have come to share our lives:** Dale F. Lott and Benjamin L. Hart, "Aggressive Domination of Cattle by Fulani

Herdsmen and Its Relation to Aggression in Fulani Culture and Personality," *Ethos* 5 (1977): 184.

THIRTEEN — The Kids in the Coal Mine

180 **still lived and worked on farms:** Dirk Johnson, "A Symbol of Rural America Fades with a Way of Life," *New York Times*, January 17, 2000.

181 **ADHD:** "ADHD for Sale," *Psychology Today*, May–June 2000, p. 17.

181 **drug most widely prescribed:** Nancy Gibbs, "The Age of Ritalin," *Time*, November 30, 1998, pp. 88–96; "Global Use of ADHD Medications Rises Dramatically," *Science Update, NIMH*, March 6, 2007.

182 **companionable:** Aaron Katcher and Gregory Wilkins, "Dialogue with Animals: Its Nature and Culture," in *The Biophilia Hypothesis,* ed. Stephen R. Kellert and Edward O. Wilson (Washington, DC: Island, 1993), pp. 173–197; Alan Beck and Aaron Katcher, *Between Pets and People* (West Lafayette, IN: Purdue University Press, 1996), pp. 143–147.

183 **zoo program success:** Aaron Katcher, Roy Erdman, Merian Waters, and the kids in the zoo program, interview by author, 1996, 2008.

185 **during stress tests:** Karen Allen et al., "Presence of Human Friends and Pet Dogs as Moderators of Autonomic Responses to Stress in Women," *Journal of Personality and Social Psychology* 61 (1991): 582–589.

186 **smaller amygdalas:** *Brain Changes Mirror Symptoms in ADHD,* Science Update, *NIMH*, July 19, 2006.

186 **delayed development:** *Brain Matures a Few Years Late in ADHD, But Follows Normal Pattern, NIMH* Press Release, November 12, 2007.

186 **very nervous rats:** Kestin Uvnas Moberg et al., "Improved Conditioned Avoidance Learning by Oxytocin Administration in High-Emotional Male Spraguc-Dawley Rats," *Regulatory Peptides* 88 (2000): 27–32.

187 **Animal trainers will tell you:** James C. Herd and Mark J. Deesing, "Genetic Effects on Horse Behavior," in *Genetics and the Behavior of Domestic Animals*, ed. Temple Grandin (San Diego: Academic, 1998), p. 219.

188 **positive social transformation:** D. M. Witt et al., "Enhanced Social Interaction in Rats Following Chronic, Centrally Infused

Oxytocin," *Pharmacology and Biochemical Behavior* 43 (1992): 855–861; Kerstin Uvnas-Moberg, *The Oxytocin Factor* (New York: Da Capo, 2003), p. 66.

190 **bursts of social competence:** Beck and Katcher, *Between Pets and People*, pp. 125–159.

190 **oxytocin in most autistic children:** Charlotte Modahl et al., "Plasma Oxytocin Levels in Autistic Children," *Society of Biological Psychiatry* 43 (1998), pp. 270–277.

190 **Clomiparamine the drug:** Margaret Altemius et al., "Changes in Neurochemistry Cerebrospinal Fluid During Treatment of Obsessive-Compulsive Disorder with Clomiparamine," *Archives of General Psychiatry* 10 (1994): 794–803.

00 **treatment of autistic patients:** Eric Hollander et al., "Oxytocin Infusion Reduces Repetitive Behaviors in Adults with Autistic and Asperger's Disorders," *Neuropsychopharmacology* 28 (2003): 193–198; Hollander et al., "Intraveneous and Intranasal Oxytocin Targets Social Cognition and Repetitive Behavior Domains in Autism: Behavioral and Functional Imaging Findings" (paper presented at the New York Academy of Sciences, March 25, 2008).

191 **sufficient supply of dopamine:** Jaak Panksepp, *Affective Neuroscience* (New York: Oxford University Press, 1998), p. 110.

191 **dopamine injections:** Crowley et al., "Excitatory and Inhibitory Dopaminergic Regulation of Oxytocin Secretion in the Lactating Rat: Evidence for Respective Mediation by D-1 And D-2 Dopamine Receptor Subtypes," *Neuroendocrinology* 53 (1991): 493–502; Kerstin Uvnas-Moberg et al., "Suggestive Evidence for a DA D3 Receptor-Mediated Increase in the Release of Oxytocin in the Male Rat," *Neuroreport* 6 (1995): 1338–1340.

191 **Noradrenalin also triggers:** E. Tribollet et al., "The Role of Central Adrenergic Receptors in the Reflex Release of Oxytocin," *Brain Research* 142 (1978): 69–84.

191 **injections of SSRIs:** Kerstin Uvnas-Moberg et al., "Oxytocin as a Possible Mediator of SSRI-Induced Antidepressant Effects," *Psychopharmacology* 142 (1999): 95–101.

191 **antipsychotic drugs are prescribed:** Kerstin Uvnas-Moberg et al., "Amperozide and Clozapine But Not Haloperidol or Raclopride Increase Secretion of Oxytocin in Rats," *Psychopharmacology* 109 (1992): 473–476.

192 **Men are better when riding:** Edward, Second Duke ofYork, *The Master of Game*, ed. W. A. and F. Baille-Grohman (London: Ballantyne, Hanson, 1904), p, 4; cited by Elizabeth A. Lawrence, "Horses in Society," in *Animals and People Sharing the World* (Hanover, NH: University Press of New England, 1988), p. 113.

FOURTEEN — Oxcytocin Deprivation

196 **collective consciousness of those of born after 1955:** Daniel Goleman, "A Cost of Modernity: Increased Depression," *New York Times*, December 8, 1992.

196 **British children consumed 68 percent more antidepressants:** "UK Leads Child Anti-Depressant Uptake," *Medical News Today,* (November 19, 2004), www.medicalnewstoday.com/articles/1659.php.

196 **16 percent of Americans:** *Harvard Science,* June 2003, http://harvardscience.harvard.edu/culture-society/articles/millions-americans-suffer-major-depression?view=print. Also see Christopher J. L. Murray and Alan D. Lopez, eds., *The Global Burden of Disease* (Cambridge, MA: Harvard University Press, 1996); Peter D. Kramer, *Against Depression* (New York: Viking, 2005); reviewed by Natalie Angier, *New York Times*, May 22, 2005.

197 **people suffering from major depression:** G. Scantamburio et al., "Plasma Oxytocin Levels and Anxiety in Patients with Major Depression," *Psychoneuroendcrinology* 32 (2007): 407–410.

197 **oxytocin deficit:** Cort A. Pedersen, "The Psychiatric Significance of Oxytocin," *Oxytocin in Maternal, Sexual, and Social Behaviors* 652 (1992): 131–148.

197 **SDD:** Sharon Heller, *Too Loud, Too Bright, Too Fast, Too Tight* (New York: HarperCollins, 2002); reviewed by Jefffrey Kluger, *Time*, November 25, 2002, p. 75.

198 **radical departure:** Wendell Berry, "Conserving Communities," *Orion,* Summer 1995, p. 49.

198 **From 1970 until 1999 Americans moved:** Roger Doyle, "By the Numbers: The U.S. Population Race," *Scientific American*, August 2000, p. 26.

199 **Educator Ann Alpert:** Katy Kelly, "False Promise," *U.S. News and World Report*, September 25, 2000, p. 54.

200 **frighteningly familiar condition:** Claudia Wallis, "Inside the Autistic Mind," *Time,* May 15, 2006. pp. 41–51.

200 **a defect in the oxytocin gene:** T. R. Insel, D. J. O'Brien, and J. F. Leckman, "Oxytocin, Vasopressin, and Autism: Is There a Connection?" *Biological Psychiatry* 45 (1990): pp. 145–157.

200 **Oxytocin deficiency has been linked to autism:** Miranda M. Lim, Isadora F. Bielsky, and Larry J. Young, "Neuropeptides and the Social Brain: Potential Rodent Models of Autism," *International Journal of Developmental Neuroscience* 23 (April–May 2005): 235–243; Insel, O'Brien, and Leckman, "Oxytocin, Vasopressin, and Autism," pp. 145–157.

200 **motherese:** Anahad O'Connor, "In Autism, New Goal Is Finding It Soon Enough to Fight It," *New York Times*, December 14, 2004.

201 **interpreting every face that looks at you as a threat:** "Eye Contact May Trigger Threatening Signals in Brains of Autistic Children," *NeuroPsychiatry Review* 6 (April 2005).

201 **stare longer:** Adam J. Guastella, Phillip B. Mitchell, and Mark R. Dadds, "Oxytocin Increases Gaze to Eye Region of Human Faces," *Biological Psychiatry* 63 (2007): 3–5.

201 **Postmortem examinations of autistic brains:** "Autistic Brain Has Fewer Neurons for Processing Emotions," *Society for Neuroscience*, July 19, 2006.

201 **look at threatening faces:** Peter Kirsch et al., "Oxytocin Modulates Neural Circuitry for Social Cognition and Fear in Humans," *Journal of Neuroscience* 25 (2005): 11489–11493; Gregor Domes et al., "Oxytocin Attenuates Amygdala Responses to Emotional Faces Regardless of Valence," *Biological Psychiatry* 62 (2007): 1187–1190; Kim M. Dalton et al., "Gaze Fixation and the Neural Circuitry of Face Processing in Autism," *Nature Neuroscience* 8 (2005): 519–526.

202 **antipsychotic drugs are prescribed:** Kerstin Uvnas-Moberg et al., "Amperozide and Clozapine But Not Haloperidol or Raclopride Increase the Secretion of Oxytocin in Rats," *Psychopharmacology* 109 (1992): 473–476.

204 **Temple Grandin has fought her way out:** personal interviews by author, 1996–2007.

204 **The pads gave me feelings of kindness:** Temple Grandin with Catherine Johnson, *Animals in Translation* (Orlando, FL: Harcourt, 2005), pp. 114–115.

205 **the way we now have our babies:** Atul Gawande, "The Score," *New Yorker,* October 9, 2006, pp. 59–67.

205 **29 percent of all babies:** Mike Stobbe, "C-Sections in US at All-Time High," Associated Press, November 16, 2005.

206 **women who deliver their babies vaginally:** Eva Nissen et al., "Different Patterns of Oxytocin, Prolactin, But Not Cortisol Release During Breast Feeding in Women Delivered by Caesarean Section or by the Vaginal Route," *Early Human Development* 45 (1996): 103–118. See also K. Uvnas-Moberg, "Neuroendocrinology of the Mother-Child Interaction," *TFM* 7 (1996): pp. 126–131.

206 **surgery and the medications it requires:** Anna-Berit Ransjo-Arvidson, "Maternal Analgesia During Labor Disturbs Newborn Behavior: Effects on Breastfeeding, Temperature, and Crying," *Birth* 28 (2001): 5–12.

206 **women who had their first elective C-section:** Miranda Hitti, "Elective Cesarean Section Deliveries Rising," *WebMD-Health,* April 22, 2005.

206 **British study found:** Susan Gilbert, "Doctors Report Rise in Elective Caesareans," *New York Times*, September 22, 1998.

207 **we are more exposed:** Andre Cicolella "We Are All Chemically Contaminated," *Le Monde*, October 12, 2005.

207 **a class of chemicals called endocrine disrupters:** Andy Coghlan, "Pollution Triggers Bizarre Behavior in Animals," *New Scientist*, September 1, 2004. Also see J. Raloff, "Common Pollutants Undermine Masculinity," *Science News,* April 3, 1999.

208 **study that links endocrine disrupters:** Miles Dean Engell et al., "Perinatal Exposure to Endocrine Disrupting Compounds Alters Behavior and Brain in the Female Pine Vole," *Neurotoxicology and Teratology* 28 (2006): 103–110.

209 **injections of oxytocin:** Maria Boccia et al., "Peripherally Administered Non-Peptide Oxytocin Antagonist L368,899 Accumulates in the Limbic Brain Areas: A New Pharmacological Tool for the Study of Social Motivation in Non-Human Primates," *Hormones and Behavior* 52 (2007): 344–541.

209 **Nasal sprays also manage:** Jan Born et al., "Sniffing Neuropeptides: A Transnasal Approach to the Human Brain," *Nature Neuroscience* 5 (2002): 514–516.

209 **most amazing molecules:** Carey Goldberg, "Feeling Shy, Afraid of Strangers? Hormone Under Study May Help," *Boston Globe*, December 26, 2005.

210 **loving looks:** G. R. Gonzaga et al., "Romantic Love and Sexual Desire in Close Relationships," *Emotion* 6 (2006): 163–179.

210 **nuturing grooming:** Kerstin Uvnas-Moberg, "The Physiological and Endocrine Effects of Social Contact," *Integrative Neurobiology of Affliation,* 1997, pp. 146–163.

210 **women who were hugged:** K. C. Light, K. M. Grewen, and J. A. Amico, "More Frequent Partner Hugs and Higher Oxytocin Levels Are Linked to Lower Blood Pressure and Heart Rate in Premenopausal Women," *Biological Psychology* 69 (2005): 5–21.

210 **Sexual Intercourse:** "Sex Before Stressful Events Keeps You Calm," *New Scientist*, January 26, 2006.

210 **reduce the excitability of each other's HPA stress systems:** A. Courtney DeVries et al., "2006 Curt P. Richter Award Winner: Social Influences on Stress Responses and Health," *Psychoneuroendocrinology* 32 (2007): 587–603.

211 **health is better:** Nicholas Bakalar, "Patterns: Exchanging Vows May Pay Off in the Long Run," *New York Times*, August 22, 2006. Also see Linda J. Waite and Maggie Gallagher, *The Case for Marriage* (New York: Broadway Books, 2000); excerpted in *Talk*, October 2000, pp. 154–155.

211 **less likely to suffer from stress:** Sarah S. Knox and Kerstin Uvnas-Moberg, "Social Isolation and Cardiovascular Disease: An Atherosclerotic Pathway," *Psychoneuroendocrinology* 23 (1998): 877–890. Also see Maria Petersson et al., "Oxytocin Causes a Long-Term Decrease of Blood Pressure in Female and Male Rats," *Physiology and Behavior* 60 (1997): 1311–1315.

211 **men benefit so dramatically:** Miklos Bodansky and Stanford L. Engel, "Oxytocin and the Lifespan of Male Rats," *Nature,* May 14, 1966, p. 751.

211 **pet ownership was the most important survival factor:** Erica Friedmann et al., "Animal Companions and One Year Survival of Patients After Discharge from a Coronary Care Unit," *Public Health Reports* 95 (1980): 307–312.

212 **Freidmann repeated the study:** E. Friedman and S. A. Thomas, "Pet Ownership, Social Support, and One-Year Survival After Acute Myocardial Infarction in the Cardiac Arrhythmia Suppression Trial," *American Journal of Cardiology* 17 (1995): 1213–1217. Also see W. P. Anderson, C. M. Reid, and G. L. Jennings, "Pet Ownership and Risk Factors for Cardiovascular Disease," *Medical Journal of Australia* 157 (1992): 298–301.

212 **Owning a cat:** E. J. Mundell, "Cats Help Shield Owners from Heart Attack," *U.S. News and World Report*, February 2, 2008, www.usnewsreport.com.

213 **put pets to the ultimate test:** Karen M. Allen, Joseph L. Izzo Jr., and Barbara E. Shykeff, "Pet Dogs or Cats, but Not ACE Inhibitor Therapy, Attenuate Blood Pressure and Renin Reactivity Among Hypertensive Stockbrokers" (paper presented at the American Heart Association Scientific Sessions, November 6, 1999).

214 **rabbits with heart disease:** Jamespaul Paredes et al., "Social Experience Influences Hypothalamic Oxytocin in the WHHL Rabbit," *Psychoneuroendocrinology* 31 (2006): 1062–1075. Also see G. L. Jennings, "Animals and Cardiovascular Health" (paper presented at the Seventh International Conference on Human-Animal Interactions, Animals, Health and Quality of Life, Geneva, Switzerland, September 6–9, 1995).

215 **cut through the isolation:** Alan Beck and Aaron Katcher, *Between Pets and People* (West Lafayette, IN: Purdue University Press, 1996), pp. 125–159.

215 **they make other people like us more:** Aubrey Fine, *Handbook on Animal-Assisted Therapy* (Burlington, MA: Academic, 2006), p. 28–29.

215 **the special look our faces take on:** Aaron Katcher and Alan Beck, "Health and Caring for Living Things," *Anthrozoos* 1 (1987): 175–177.

216 **children today are more anxious:** Christine Gorman, "Stressed-Out Kids," *Time*, December 25, 2000, p. 168.

216 **until people feel safe and connected:** Susan Griffith, "The Age of Anxiety," *Case Western Reserve University Magazine,* Spring 2001, p. 36.

217 **epitaph for Botswain:** Poem printed in *TLS*, June 6, 2008.

217 **only when partners were loving:** Sharon Lerner, "Good and Bad Marriage, Boon and Bane to Health," *New York Times,* October 22, 2002.

FIFTEEN ⁓ Just Realizing

222 **their urge to keep a pet:** Humane Society of the United States, 2007–2008.

222 **generations to suppress:** Robert B. Edgerton, *The Individual in Cultural Adaptation* (Berkley: University of California Press, 1971), p. 254. Also see Cynthia Graber, "ADHD Genetics Sometimes Beneficial," *Scientific American*, June 13, 2008.

223 **6 billion hungry mouths to feed:** Julie Lewis, "Six Billion and Counting," *Scientific American,* October 2000, p. 30.

223 **agritainment:** Anita Hamilton, "That's Agritainment," *Time*, October 31, 2005, p. 72.

225 **back to the garden:** Anne Raver, "Tutored by the Great Outdoors," *New York Times*, October 7, 1999.

225 **narrowing of consciousness:** John Lahr, "Past Forgetting," *New Yorker,* May 15, 2000, p. 88.

226 **canary:** Roger Welsch, "A Song for the Pioneers," *Audubon,* November 1992, pp. 112–115.

227 **dog seemed to know:** "An Epileptic Child's Best Friend," *Washington Post*, June 22, 2004.

227 **a potent antioxidant:** S. O. Iseri et al., "Oxytocin Protects Against Sepsis-Induced Multiple Organ Damage: Role of Neutrophils," *Journal of Surgical Research* 126 (2005): 73–81. Also see Bernd Moosman, and Christian Behl, "Secretory Peptide Hormones Are Biochemical Antioxidants," *Molecular Pharmacology* 61 (2002): 260–268.

228 **therapy dogs in hospitals:** Lawrence K. Altman, "Study Identifies Heart Patient's Best Friend," *New York Times*, November 16, 2005.

228 **When we see a smiling face:** Jane E. Warren et al., "Positive Emotions Preferentially Engage an Auditory-Motor 'Mirror' System," *Journal of Neuroscience* 26 (2006): 13067–13075. Also see Marco Iacoboni, "Mental Mirrors," *Natural History,* May 2008, p. 37. Also see Paul Ekman and Richard J. Davidson, "Voluntary Smiling Changes Regional Brain Activity," *Pscychological Science* 4 (1993): 342–345.

229 **our pets make us laugh:** Patricia Long, "Laugh and Be Well?" *Psychology Today*, October 1987.

230 **watching nature films:** Lennart Levi, "A Biopsychosocial Approach to Etiology and Pathogenesis," *Acta Pharmacologica Scandinavica Supplement* 160 (1997): 103–106.

230 **watching nature films:** Alan Beck and Aaron Katcher, *Between Pets and People* (West Lafayette, IN: Purdue University Press, 1996), p. 115.

231 **elephant talk:** Katherine Payne, "Elephant Talk," *National Geographic,* August 1989, pp. 264–277.

231 **language talents in vervet monkeys:** Dorothy L. Cheney and Robert M. Seyfarth, *How Monkeys See the World* (Chicago: University of Chicago Press, 1990), pp. 139–175.

233 **Great Ape Trust:** www.greatapetrust.org. Also see TED Talks: Susan Savage-Rumbagh: Apes that Write, Start Fires, and play Pac-Man, www.ted.com/index.php/talks/susan_savage_rumbagh_on_apes_that_write.html.

234 **Elk are one of the breeds:** Edward Hymans, *Animals in the Service of Man* (London: Dent, 1972), p. 149.

234 **are now being bred:** T. Williams, "The Elk Ranch Boom," *Audubon,* May 1992, p. 14.

235 **Pliny the Elder:** J. M. C. Toynbee, *Animals in Roman Life and Art* (Ithaca, NY: Cornell University Press, 1968), pp. 206–207.

235 **Dolphins off the shore of western Australia:** www.monkeymiadolphins.org. Also see Horace Dobbs, *The Magic of Dolphins* (Cambridge, UK: Lutterworth, 1990).

235 **In Ireland's Dingle Bay:** www.dodingle.com/pages/fungi dingle dolphin.html.

236 **just watching dolphins is not enough:** Leatherwood and Ackerman, "How Can You Tell It's Love?" pp. 210–213.

237 **marine mammal program:** Don Oldenburg, "The Navy's Dolphin Safe Program," *Washington Post*, April 7, 2003.

239 **more than 120 million people:** Michael Cannell, "Ice Age at the Zoo," *Washington Post Magazine*, October 10, 1999.

240 **Madagascar Fauna Group:** www.savethelemur.org.

240 **creating egg and sperm databases:** Robert P. Lanza, Betsy L. Dresser, and Phillip Damiani, "Cloning Noah's Ark," *Scientific American,* November 2000, pp. 84–89. Also see Jared Diamond, "Playing God at the Zoo," *Discover,* March 1995, pp. 79–85.

240 **28 percent of the African continent:** *Wilderness: Earth's Last Wild Places,* ed. R. A. Mittermeier et al. (Chicago: University of Chicago Press, 2003).

240 **Memories of Papua New Guinea's jungles:** Andrew C. Revkin, "Forests Disappearing in Papua New Guinea," *New York Times,* June 3, 2008.

241 **Pharming:** Sheryl Gay Stolberg, "Could This Pig Save Your Life?" *New York Times Magazine*, October 3, 1999, pp. 46–51. Also see Michael O'Donnell, "Pork Futures," *International Management,* April 1993, p. 63; Albert Rosenfeld, "New Breeds Down on the Pharm," *Smithsonian Magazine,* July 1998, pp. 22–30; William H. Velander, Henryk Lubon, and William N. Drohan, "Transgenic Livestock as Drug Factories," *Scientific American,* January 1997, pp. 70–74.

242 **most valued member of the family:** Karen Allen, "Companion Animals and Elderly People," in *Healthy Pleasure of Their Company*, www.deltasociety.org.

242 **closer to their pet:** Sandra B. Barker, "Therapeutic Aspects of the Human-Companion Animal Interaction," *Psychiatric Times* 16 (February 1999).

243 **benefits of emotional generosity:** John O'Neil, "Help Others for a Longer Life," *New York Times*, November 12, 2002.

243 **care and feeding of animals:** *Time*, February 4, 2008, p. 18.

244 **to witness mass extinctions:** Mark Ridely, "The New Age of Extinction," *Times Literary Supplement,* August 13, 1993. Review of E. O. Wilson, *The Diversity of Life* (London: Penguin, 1993).

Acknowledgments

I am grateful to the many humans and animals who helped to make this book possible. I'd like to thank Stephanie Powers and Mick Kazarowski, who first challenged me to think hard about the history of human-animal bond. Andrew Rowan opened up his rolodex (back in the day) and pointed me to Aaron Katcher and so many others who provided pieces to this amazing puzzle. And a special thank-you to Sue Carter, Kerstin Uvnas-Moberg, and Cort Pedersen, who took me seriously and guided my search for the biology of the human-animal bond. Thank-you Ray Hunt for showing me that biology at work. Temple Grandin helped me see the bond from an animal's point of view. And where would I be without the great Barbaras, Thorne and Smuts, who thought I got it right and wanted everyone to know. Edward O. Wilson's praise and support have made me prouder than I can say.

Thanks to Cindy D'Agostino and Sandy Mack, for your enthusiasm and welcome edits. And to Sydell Herbst, my angel from Brooklyn, who said I was "the messenger," which is what it feels like. Thank-you simply won't do it for my other (arch) angel, Michael, who also insisted I was born to tell this story, and to the elegant Merloyd Lawrence who believed in and bettered this story into a book.

And to the many other friends and family who listened patiently and cheered me on, thanks, I needed that. And to my sister Ginny for capturing the essence of the human-animal bond in her perfect illustrations.

But there would be no words, or story, or book if some very special animals hadn't made sure I felt both the biology and the bond. They know their names, and they already know what they've meant to this book and to my life. But I thank them anyway, every minute of every day.

And finally I'd like to remember and pay tribute to two great observers of the human-animal bond, the veterinarian and anthropologist Elizabeth Lawrence and the biologist Dale Lott who inspired and supported this effort. They left this world a better place for all humans and animals.

—*Meg Daley Olmert*

Index

Aboriginal societies, 143, 158
Abydos, 152
Acceptance, 174–175
ACE-inhibitor drugs, 213
Adaptation, 42, 68
Adrenalin, 49, 230
Adrenocorticotrophic hormone (ACTH), 49
Affection, 71, 189
Age of Domestication, 149
Aggression, 48, 85, 129, 173, 174, 178, 183
  breeding out, 155
  intimacy and, 175
  oxytocin and, 21
Agritainment, 223–224
AIDS, 241, 242
Alexander the Great, 84
Alleles, 24, 62, 63, 78, 233
Allen, Karen, 185, 188, 212, 213
Alloparents, 56, 59, 142
Alpert, Ann, 199
Alzheimers, animals and, 189
Amygdala, 47, 48, 57, 186, 201, 210
Anatomy, 5, 34–35
Anderson, B. K., 98
Anger, oxytocin and, 29
Animal husbandry, 136, 156, 167
Animals
  attachment to, 175, 243
  caring for, x, 152, 153–154, 198
  disappearance of, 220, 244
  early depictions of, 64–65
  encounters with, 53, 230, 239
  fascination with, viii, 16, 175, 239
  hand-feeding, 78–79
  impact of, 184, 193
  imprint by, 9
  merging with, vii, 232–233
  painting, 36, 37
  relationship with, 11, 163, 185–186, 199, 216–217, 240–241
  respect for, 175
  as superorganisms, 5
  understanding, 82, 172
Antianxiety medications, 196
Antibiotics, 205
Antidepressants, 191, 196, 202, 209
Antioxidants, 227–228
Antipsychotics, 191, 196, 202, 209
Anxiety, 28, 92, 185, 187
  oxytocin and, 56, 216
Apes, 56, 231, 233
Arthritis, 228
Artificial limbs, 102
Asperger's syndrome, 200
Associative learning, 159
Atherosclerosis, 214, 228
Atrazine, 207
Attention, 58, 119, 124
Attention deficit hyperactivity disorder (ADHD), ix, 181, 186, 191, 192
  social ineptness and, 187–188
  zoo program and, 182–183
Attitudes, interprcting, 130
Attraction, 31, 59
Aurochs, 144, 145, 159
  domestication of, 155, 167
  taming of, 147, 148–149
Autism, 36, 37, 201, 203

Autism *(continued)*
animals and, 189–190
comfort for, 204
oxytocin and, 112, 190–191, 200, 202, 210

Baboons, 161, 242
communication with, 114–118
social agenda of, 115–116
Baby-sitters, 54, 56, 59, 61
Badgers, coyotes and, 69–70, 220
Bales, Karen, 21
Barasana women, 72
Barking, 127, 128, 131
Barnett, John, 156
Bartels, Andreas, 30
Bastet, 161
Bateson, Gregory, 112
Beard, George, 100
Behavior, 5, 93, 116, 134, 173, 175
affiliate, 45
animal, 10, 78, 122, 187, 188, 231
brain and, 17
computers and, 199
controlling, xii, 163
emotional, 208
learning, 128
maternal, 18, 21, 30, 57, 58, 59
observing, 34–35
oxytocin and, xiv, 23, 30, 28, 46
parental, 22, 54, 130
repetitive, 191, 200
reproductive, 207
shifts in, 53, 191
social, 23, 25, 44, 129, 191, 196, 207
traits/passing on, 57
understanding, 38, 39, 130
vasopressin and, 23
vocal, 128
Behavioral disorders, ix, 181, 191, 192, 196
Belyaev, Dmitry, 75–78, 128, 134, 155
Beta endorphins, 50, 74, 191
Bioethics, 163
Biophilia, 12–13, 32, 44
mind/body moment of, 13
neurochemical capacity of, 221–222
oxytocin and, 82, 103
Biotechnology, 163, 241–242
Bipolar disorder, 197
Blood pressure, 185, 213, 227
lowering, xiii, 73
oxytocin and, 214
pets and, 212
Bloom, Paul, 126
Boatswain (dog), epitaph for, 217
Boccia, Maria, 58
Body language, 91, 116, 129, 191
Bonding, 46, 82, 85, 158, 159, 233, 237
age of, 162, 171
biology of, 19, 24, 72, 77, 162–163
breaking, 180, 198
chemistry, 56
DNA and, 24
human-animal, ix, xi, xiii, xvi xvii, 2, 14, 24, 25, 31, 84, 97, 107, 156, 164, 230, 232, 240, 241, 242, 243–244
maternal, 20, 112
oxytocin and, xiii, 21, 22, 23, 26, 31, 93, 169, 175, 188–189
parental, xi, xii
social, x, xii, 19, 22, 25, 31, 50, 62, 103, 174, 207, 216
vasopressin and, 23
Border collies, communication with, 124–126

Botai, 83
Brain, xvi, 4–5, 35, 119, 192
abnormalities, 186
autistic, 201
behavior and, 17
emotion and, 17
evolution of, 186
growth of, 3, 4, 8, 9, 11, 14, 54
language and, 10
movement and, 5
nonconscious part of, 39, 52, 101
nonlinguistic, 108
oxytocin and, 17
reputation and, 153
social, 3, 22, 47, 62, 152, 153
subconscious, 6
vocal tract and, 106
wiring, 7, 9
Brain cells, 5, 6, 49
Brain stem, oxytocin and, 26
Breast-feeding, 57, 58, 72, 74, 157, 158
oxytocin and, 29, 30, 46, 131, 187, 188
vasopressin and, 131
Breast milk, 72
making/delivering, 44
oxytocin and, 26, 206
Breath, 107, 244
Breeding, 136, 149, 152, 155, 159, 161, 167, 182, 234
captive, 240
selective, 156, 238, 240
Byron, Lord: dog of, 217

Cahill, James, 141
California Department of Education, 224
Call, Josep, 124–125
Calm/connect, state of, 50, 92
Calming, 75, 95, 154, 185, 204, 214, 216, 234
oxytocin and, 19, 73
selective breeding for, 238
Camels, 175, 176, 177
Cardiovascular disease, 211, 213
Caring, xv, 152, 153–154, 157, 159, 169, 198, 222, 227, 243, 244
oxytocin and, 217, 219
Carnivores, 122, 221
Carpenter, William, 99
Carter, C. Sue, x–xi, xii, 20, 21, 22, 91
Catalhoyuk, 143–147, 149, 161
Cats, 145–146, 216
burial rights for, 161
secret language of, xv–xvi
well-being and, 160–161, 212–213
Cattle, 156, 169, 172
behavioral properties of, 175
breeding, 155, 159, 167
comforting, 204
communicating with, 170, 175
domination of, 178
environment and, 156
fear and, 203
grooming, 168, 174
hysterical, 203–204
milking, 155, 159
oxytocin and, 149
social dominance and, 173–174
socializing, 175
status and, 167–168
taming of, 148
Cave paintings, 34–35, 37, 64–65, 69
Center for Behavioral Neuroscience, 23
Cerebrospinal fluid (CSF), 18
Cesarean section, 205, 206, 207
Champagne, Frances, 58
Cheney, Dorothy, 113, 232

Childbirth, 18, 55, 205, 207
 oxytocin and, 197, 206
Civilization, xiii, 82, 180, 197, 221
Clark, Kenneth, 161
Clever Hans (horse), 88, 89, 90–91, 100, 110–111, 113
Clomipramine, 190
Clottes, Jean, 34–35
Clutton-Brock, Juliet, 83, 129–130, 155
Cognition, 19, 187
Coleman, Graham, 156
Comfort, 91, 202, 204, 221
Commitment, 31, 62, 135
Communication, 122, 131, 227
 baboon, 116–117
 human-animal, 97–99, 112–113, 123, 238
 networks, 9
 nonverbal, x, 93, 108, 129
 oxytocin and, 56
 paralinguistic, 112
 social, x, 129
 social abilities and, 126
 vocalization and, 106, 114
Communities, 139, 144, 198
Computers, 199, 231
Concentration, 40, 101
Connections, 16, 60, 223, 241
Consciousness, 7, 37, 39, 71, 162
 animals and, 12
 language and, 8
 narrowing of, 225
 primitive, 52
Contentment, 118, 119, 154, 170, 221, 222, 223
Cooperation, 3, 46, 70, 97, 99, 139, 154, 220–221, 234
 social, 64
 among wolves, 66, 69
Cort, 75, 77
Corticosterone, 49
Corticotrophin-releasing hormone (CRH), 48–49
Cortisol, 49
Cow dance, 170, 171
Cowboys, as hybrid apart, 178
Cows. *See* Cattle
Coyotes, 216
 badgers and, 69–70, 220
Culture, 11, 65, 162
 cattle, 170, 171, 172, 174
 interspecies, 220
 language-based, 233
 pan-homo, 233–234
 subsistence, 135

Darwin, Charles, 112
Davidson, Richard, 201
DDT, 207
Depression, 196, 197, 210, 214–215
Developmental disorders, 36, 181, 200
Devereux Foundation, ix, x, 181, 184, 193, 196
 zoo program and, 188, 225
Devotion, xv, xvi, 14, 189
Diabetes, 200, 212, 228
Diamond, Jared, 40, 41, 158–159
Diethylstilbestrol (DES), 208
Dinka, cattle and, 170, 171
DNA, 12, 19, 25, 62, 75, 78, 163, 231, 240–241
 bonding and, 24
 merging, 242
Dogs, 222, 235
 antioxidants and, 228
 domestication of, 145
 foxes and, 238
 heart protection and, 212, 213
 herding by, 142
 humans and, 122, 136
 keeping, 138
 muscle reading by, 122
 music and, 123
 oxytocin and, 123, 211–212

paralanguage and, 122
physical/spiritual well-being and, 135
psychology/behavior of, 134
pushing/dragging by, 132
superior senses of, 133–134
therapy, 228
wolves and, xiv, 17, 71, 130, 131, 238
Dolphins, 231, 237–238
domestication of, 235–236
Domesticated elite, 78, 238
Domestication, xiii, xiv, 66, 68, 75, 76, 77, 84, 86, 145, 147, 155, 160, 161, 167, 169, 176, 180, 197, 223–224, 229, 230, 234, 235–236
biotechnology and, 163
breeding/feeding, 152
demands of, 159
downside of, 157
elements of, 149
history of, 158–159
intervention and, 142–143
oxytocin and, xiii, 157
social arrangements for, 106
taming and, 159
vestiges of, 224
women and, 149
Dominance, 103, 173–175, 178, 208
Dopamine, 49, 50, 74, 191
Down syndrome, 200
Dudley, Susan, 141
Durbeyfield, Tess, 157, 215, 223, 224

Egyptians, 156, 162, 167
artwork of, 157–158
cats and, 160–161
domestication and, 160, 161
Fulani and, 171
Nubians and, 166, 168
taming by, 158
Ekman, Paul, 108–109
Elephants, 176, 231, 232
Elk, riding, 234
Emotions, 16, 72, 82, 91, 109, 116, 124, 162, 163, 177, 189, 201, 208, 217, 220
brain and, 17
controlling, xii
guiding, 13–14, 31, 131
intentions and, 107
oxytocin and, xii, 32, 112
pleasing, 79
social life and, 31
subsistence cultures and, 135
Endocrine disruptors, 207, 208
Engell, Miles Dean, 208
Environmental challenges, 26, 69, 152, 192
Equitation, 84, 96–97, 102–103
Estrogen, 25, 207
oxytocin and, 73, 111, 149
Evans, Nicholas, 85
Evans-Pritchard, Sir Edward, 168, 169, 170
Evolution, 9, 46, 61, 117

Facial expressions, 9, 10, 56, 106, 107, 109, 116, 129, 201, 203
Farmers, 180, 198, 222, 232
cats and, 160
hunters and, 147, 163
oxytocin and, 223
Farming, 140–141, 142, 143, 147, 164, 223
Farms, 222, 223–224
leaving, 180, 198
Fear, 48, 52, 57, 92, 203
oxytocin and, xiv, 47
Feedback system, 49–50, 95, 157
Feeding, 148, 152, 159, 182, 243
Feh, Claudia, 94
Fehr, Ernst, 60–61
Feinstein, Mark, 127, 128

Fight/flight response, xii, xiv, 47, 48, 49, 50, 82, 93
 chemistry of, 53
 oxytocin and, 92, 225
File, Amanda, 141
Fischer, Julia, 124–125
Fondon, John, 77
Foré tribe, 41–42, 52, 109, 240
Foxes, 75–76, 238
Francis, Darlene, 58
Friedman, Erica, 212
Friendship, xvi, 128, 131
Fukuyama, Francis, 163
Fulani, 177–178, 180
 cattle and, 171–172, 173, 175
 dominance/acceptance and, 174–175
 herding by, 171–174
Fungie (dolphin), 236

Gardens, 153, 225
Garner, Harold, 78
Generosity, 60, 243
Gestures, 124, 201
 hand, 9–10, 203, 238
Gladwell, Malcolm, 39
Grandin, Temple, 202, 203, 204–205
Grazing, 140, 142, 159
Great Ape Trust of Iowa, 233
Grooming, 94, 148, 168, 173, 174, 182
 mutual, 85
 oxytocin and, 227
Growls, 112, 119
Grunts, xv, 112, 113, 116, 232

Hammock, Elizabeth, 24, 25
Hardy, Thomas, 157, 215
Hare, dogs and, 131–132, 133, 134, 135, 136
Hare, Brian, 122, 123
Harrison, Mary Scott, 225
Hart, Benjamin, 175, 178
Heart health, 212, 213
 oxytocin and, 211, 214
Heart rate, 185, 212
 lowering, 73, 74, 94
 oxytocin and, xiii
Heartsick, 214, 215
Heller, Sharon, 197
Hemsworth, Paul, 156
Herders, 132, 136, 173, 180, 232
Herding, 142, 143, 145, 170–173
Herds, 103, 166, 178
Herodotus, 96
Hidatsa, 141, 142
Hobhouse, John Cam, 217
Hodder, Ian, 144–145, 146
Hollander, Eric, 112, 190–191, 201
Homo sapiens, 8, 9
Homonids, 2, 53, 54
Horsemanship, 84–85, 104
Horses
 communication with, 85, 86, 93, 104
 disappearance of, 83
 domestication of, 86
 domination of, 178
 oxytocin and, 110–111
 as prey animals, 84
 relationships with, 85, 87, 98
 touching, 93–94
 vision of, 88, 91, 110
Horse's feet, knowing, 98, 99, 102
Horse's idea, blending with, 86, 101–102
Hrdy, Sarah Blaffer, 55, 56, 59
Human-animal bond, ix, xiii, 2, 24, 25, 97, 107, 156, 164, 230, 232, 240, 241, 242, 243–244
 biology, xvi–xvii
 chemistry of, 14
 oxytocin and, xi, 31
 successful, 84

Humans, 229–230
aurochs and, 159
canines and, 122
cats and, 160–161
cows and, 156
dogs and, 74, 136
horses and, 91
wolves and, 66, 70, 75, 78, 106, 135–136, 159
Humphrey, Nicholas, 36–37
Hunger, 16, 44
Hunt, Ray, 109, 148
body language of, 91
communication by, 84, 86, 87–88, 96, 97–98
horse's feet and, 99, 102
muscle reading by, 101
touch by, 93–94
visualization and, 102
Hunter-gatherers, 40, 143, 221
Hunters, 9, 130, 132, 139, 144, 148, 169, 170
comfort for, 136
farmers and, 163
hunted and, 3
self-esteem of, 52
Hunter's trance, 13, 30, 115
Hunting, 4, 69, 145, 152, 155
alliances for, 8
big-game, 63, 64
culture/consciousness and, 169
giving up, 222
horses and, 83
lodges, 144
Hyperactivity
animals and, 184
endocrine disruptors and, 207
Hyperavoidance, 201
Hypermetabolism, 28
Hyperscrutiny, 38
Hypersensitivity, 202
Hypothalamus, 17, 25–26, 27, 48, 95, 214
Ice Age, 2, 31, 52, 139, 144
behavior shifts in, 53
end of, 138
first dog owners of, 127
neurohormonal experiments of, 75
sensory experiences of, 8
stress of, xiv
Ice Age babyphilia, rise in, 59
Ice Age children, 54, 56, 58, 64
Ice Age fathers, paternal personality of, 62
Ice Age mothers
attention from, 58
oxytocin and, 55
parental obligation for, 54
personality shift for, 55
trust and, 59
Identities, 32, 103, 163
Ideomotor actions, 99–100, 100–101, 103
Imagination, 106, 234
Immune system, oxytocin and, 228
Industrial revolution, 196, 198
Insel, Thomas, 22, 23, 210
Intentions, 107, 124, 173
nonthreatening, 69
Interaction, 49, 91, 92, 180
oxytocin and, 29, 243
Interspecies bargains, 159–160
Interspecies codependence, 69
Interspecies enterprise, 166
Interspecies merging, 241
Interspecies understanding, 14
Intimacy, 32, 175
Inuit, dogs and, 132
Isolation, 225, 226

Jacobson, Edmund, 99–100
James, William, 100
Jaynes, Julian, 35, 38, 42
Jerison, Harry, 3, 7–8, 12, 68
Julius Caesar, 147–148, 155
Just watching, 5–8, 14, 30, 32, 34

Kalahari Bushmen, 143
Kalam people, 41
Kamen, Dean, 102
Kaminski, Juliane, 124–125
Karl XI of Sweden, elk and, 234
Katcher, Aaron, ix–x, xi, 189
 ADHD and, 181
 on farm/factory, 180
 hyperactive children/pets and, 184
 on pets/relaxation, 215
 zoo program and, 183, 190, 192
Kellert, Stephen, 10
Kendrick, Keith, 18
Keverne, E. Barry, 18
Kinship, xiv, 32, 122, 143
Klein, Richard, 34
Kohler-Rollefson, Ilse, 177

Labor
 deaths during, 56
 oxytocin and, xii, 16–17, 18, 19, 31, 61, 205
Lactation, 44, 208
 oxytocin and, xii, 16–17, 27–28, 30, 31, 45, 61
Language, 7, 36, 37, 106, 162, 190
 animals and, xv–xvi, 11, 12
 ape, 233
 articulate, 107–108
 body, 91, 116, 129, 191
 brain and, 10
 canine, 124–125
 consciousness and, 8
 developing, 11, 40, 126
 learning, 128
 social, x, 123
 social communication and, 129
 spoken/written, 107, 108
 symbolic, 231
Lawrence, Elizabeth, 11, 103
Lear, King, 79
Levant, 138–139, 140, 149
Lévi-Strauss, Claude, 128
Lewis, Charles, 153
Libet, Benjamin, 100–101
Licking, 57–58, 62, 173, 174, 227
Lienhardt, Godfrey, 170
Linguistics, 126, 129
Linnaeus, 44
Lisinopril, 213
Livestock, 145
 caring for, 157, 169, 203, 222
Lott, Dale, 78–79, 175, 178
Luke arm, 102

Macaque monkeys, endocrine disrupters and, 208
Madagascar Fauna Group, 240
Mammals, 44, 46
Mammary nerves, oxytocin and, 74
Marine mammal program, 237
Marriage, 211, 217
Mating, 19–20, 22
McCune, Lorraine, 114, 127
Meaney, Michael, 57, 58
Meditation, 83, 216
Meintjes, Roy A., xiii, 74, 191, 213
Mellaart, James, 146, 149
Mental skills, 35, 41
Mere exposure effect, 6, 230
Metaphor, 11, 134
Methoxychlor (MXC), 208
Midgely, Mary, 108, 112, 130
Midwifery, oxytocin and, 156
Migration, 198, 222
Milk, 167, 173
Milking, 155, 157, 159, 215
Mind, 7, 38, 109
Mind reading, 111, 201
Minnesota Stroke Research Center, 213
Mirror neurons, 5–6, 101
Mithen, Stephen, 38, 82
Modahl, Charles, 37
Mongols, horses and, 84
Monkeys, 113, 208, 231–232
 talking by, 232

Monogamy, xi, 22
Morton, Eugene, 106–107, 112
Motherese, 123, 200, 229
Motion, 100, 102
Motor skills, 2, 101
Movement, 5, 10, 101, 129
Mumford, Lewis, 149
Murals, 144, 145
Muscle reading, 100, 101, 103, 110, 111, 122
Mutual parasitism, 168, 169

Nadia (autistic girl), 36–37, 42
Nagasawa, Miho, 126
National Institute for Psychological Sciences, 152
National Institutes of Mental Health, 22, 152, 186, 210
Naturalism, 36
Nature, xvi
- affiliation with, 11
- changing nature of, 154–155
- inspiration/cues from, 34
- laws of, 135
- returning to, 199, 225

Neanderthals, 9
Neolithic Age, 40, 146, 159, 160, 180
Neolithic society, 143, 144, 145, 152, 153, 154, 155, 167
Nerve, 26, 53, 54
Nerve disorders, 210
Nerve signals, 102
Nervous system, 44, 63, 133–134, 199
Neural network, 26, 47, 49, 107, 200
Neurobiology, 65, 66, 72, 228
Neurochemistry, 47, 59–60, 139, 156
Neuroeconomics, 59, 60, 63, 79
Neurohormones, 75, 106, 211
Neurotransmitters, 48, 74
Nonnoxious stimulation, 185, 205
Noradrenalin, 48, 57, 191, 230
Nubians, animals and, 166, 167, 168, 170
Nuer, 168–169, 170
Nurturing, xv, 18, 59, 71, 72, 76, 138, 149, 153, 154, 166, 167, 169–170, 174, 180, 188, 202, 205, 220

Observation, xiv, 86
Odendaal, Johannes, xiii, 74, 191, 213
Olfactory system, 3, 7
Opioid receptors, 94
Organ transplants, interspecies, 242
Ouija boards, 99, 102
Oxytocin
- decrease in, 206
- deprivation, 193, 196–197, 200
- early infusion of, 59
- estrogen and, 73, 111, 149
- exchange, 30
- exposure to, 186, 187
- heritage, 32
- impact of, xiv, 17–18, 19, 22, 30, 47, 54–55, 169–170, 191–192, 227–228
- increase in, xiii, 25, 61, 74, 126, 154, 185–186, 191, 192,196, 201, 202, 210
- injections of, 186, 209
- precursor versions of, 46
- receptors, 22, 23, 47, 58, 209
- release of, xi–xii, xiv–xv, 74, 92, 94, 95–96, 97, 99, 154, 174, 188, 189, 191, 214
- synthetic, 205
- tampering with, 207
- vasopressin and, 21–22, 63, 65, 129, 175, 207

Oxytocin-blocking drugs, 21, 73

Packer, Craig, 34–35
Pain, 92, 93, 114, 225
removing, 205, 227
tolerance of, 154
Paleolithic, 61, 65, 144
Pallas, 4, 52
Papua New Guinea, 40, 240
Paralanguage, 107, 112, 119, 122, 123, 124, 129, 130, 203
Parasympathetic activity, 27, 128
Parentese, 123
Parkinson's disease, 49
Partnerships, xiv, 19–20, 99
Pastoralism, 157, 174
Pastoralists, 163, 169, 170, 221
Paterson-Brown, Sara, 206
Pedersen, Cort, xii, 17, 18, 19, 58, 59
Personality, 28, 54, 163
Pets, 160, 185, 189, 190
bond with, 14
caring for, 14, 217
hyperactive children and, 184
love for, 229, 242–243
oxytocin and, xiii, 211–212, 213, 214, 215
quality/quantity of life and, 228
relaxation and, 215
wolves as, 127
Pfungst, Oskar, 89, 90–91
Pharming, 241
Physical vocabulary, oxytocin/vasopressin and, 208
Physiological factors, 28, 45, 57, 74
Pitocin, 205, 209
Pituitary glands, 17, 49
Plants, 221
caring for, 152, 153, 198
social recognition for, 141
Play-bow, 129
Pleistocene, 2, 53, 61, 63
Pliny the Elder, 235
Posture, 54, 58, 116, 129
Powers, Stephanie, ix
Prairie voles, xi, 20, 23, 62, 208
Prange, Arthur, Jr., 17, 19
Predators, 2, 4, 52, 63, 69, 82, 83, 85, 127, 142, 232
ascent of, 53
avoiding, 207
prey and, 3, 6, 10
Pregnancy, 25, 208
Prey, 84
ascent from, 53
predators and, 3, 6, 10
Princess (cat), story about, xv–xvi
Prongracz, Peter, 128
Protolanguage, 8, 130
Psychological changes, 28, 163, 189
Psychological studies, 196–197

Raika, 176–178, 180
Rajasthan, camel and, 176
Recognition, 18, 31, 62, 106
social, 22, 23, 47, 103, 141, 143, 155, 159
Red deer, breeding/bottle-feeding, 234
Relaxation, xiv, 50, 64, 86, 91, 96, 110, 143, 180, 183, 205, 220, 229
oxytocin and, 45
pets and, 215
stroking and, 73
Reproduction, 50, 207
oxytocin and, 45, 46
Reputation, 152, 153
Responsibility, 27, 29, 61, 77
Rico (dog), communication with, 124–125
Riding, ix, 83, 85, 96–97, 192, 234
bareback, 97, 98
Ring, Robert, 209–210
Ritalin, 181, 191
Rituals, 138, 144, 145, 146, 147, 174, 178

Rizzolatti, Dr., 5–6
Roaming, 143, 222
Roth, Philip, 225
Roth, W. E., 72

Safety, 3, 44, 127, 183, 187
St. Roche, licking dog and, 227
Salt, taming and, 148
Savage-Rumbaugh, Susan, 233–234
Savants, 36–38, 42, 200
Savishinsky, Joel S., 133, 134
Scavenging, 4, 70
Schizophrenia, oxytocin and, 210
Scythian nomads, horses and, 96
Self-esteem, 52, 79, 183
Self-image, 64, 79
Self-interest, 58, 62, 198
Sensory defensive disorders (SDD), 197
Sensory deprivation, 226
September Commission, 88
Serotonin, 50, 77, 191
Serpell, James, 72
Sexual intimacy, 20, 45, 50, 63
  oxytocin and, 208, 210
Seyfarth, Robert, 113, 232
Shakespeare, William, 79
Shamans, 144
Shepard, Paul, 3, 30
Shiva, Lord, 176
Siberian silver fox, 75
Silent seeing, art of, 40
Skin, oxytocin and, 93
Sled teams, human-dog, 134
Sleep, 45, 47, 127
Small, Meredith, 54
Smuts, Barbara, 114–115, 116–118, 119
Sociability, 62, 70, 190
Social agenda, 47, 115–116
Social alliances, 64, 85
Social behavior, 44, 129, 191, 196
  adjusting, 25
  chemistry of, 207
  differentiation on, 25
  promoting, 23
Social bonds, x, xii, 19, 62, 103, 174
  formation of, 22, 31, 207
  oxytocin and, 25, 216
  satisfying/enduring, 50
Social boundaries, establishing, 208
Social carnivores, 2, 16, 65, 68
Social chemistry, communities and, 139
Social code, oxytocin and, 45
Social competence, 190, 202
Social connections, 32, 112, 181–182, 188
Social contact, 28, 50, 189, 202, 205
Social contract, 49, 64, 180, 221
Social deprivation, 226
Social desirability, 55, 188, 229
Social drive, increase in, 92
Social engagement, 114–115, 216
Social experiments, 154, 197
Social imagination, 59, 71, 139
Social intelligence, 59, 61, 203
Social interaction, 44–45, 91, 175, 185
  mammalian growth and, 47
  oxytocin and, 45, 154, 180
Social mammals, 97, 226–227
Social recognition, 22, 47, 141, 143, 155, 159
  oxytocin and, 23, 103
  vasopressin and, 103
Social risk, 61, 199
Social status, brain and, 152, 153
Socializing, xiv, 23–24, 75
Sound, hypersensitivity to, 202
Sound-meaning correspondence, 114, 126
Southern Pines Elementary, 225
Spanish Harlem, gardens in, 153
Species barriers, crossing, 72, 149
Squeeze chute, 203–204

Starvation, 83, 132
Stimulation, 6, 196, 230
Stress, xiv
coping with, 58, 207
HPA, 48, 49, 57, 197, 210–211
increase in, 91–92
oxytocin and, 27, 209, 210, 211, 213, 214, 217
reducing, 229
social, 152
tests, 213–214
Stress hormones, xi, xiii, 27, 48, 72, 75
lowering, 74, 187
Stroking, 73, 94, 95, 154, 174
Stumpf, Carl, 88
Submission, 148, 159, 172
Subsistence cultures, emotional displays and, 135
Survival strategy, 5, 122, 127, 220
Sympathetic nerve action, decrease in, 128

Tameness, 76, 77, 128
Taming, 72, 148–149, 158, 221, 230, 235
domestication and, 159
oxytocin and, 149, 210, 239
Tarpan, 83
Technology, 82, 95, 101, 140, 198–199, 231
Testosterone, 207, 208, 211
Teu, 41
Therapy, xv, 228
animal, 74, 183–184, 202
sensory, 202
touch, 192, 202
Thought, 100, 106
Tomb paintings, 158, 167, 171
Touch, xv, 72, 96, 107, 158, 192, 196, 201, 202, 204, 222, 227
casual, 50
hypersensitivity to, 202
oxytocin and, xi, 73, 93–94, 95, 206
plants and, 141
Toynbee, J. M. C., 147
Transformation, 25, 75, 188
Transhumance, 198
Trevathan, Wendy, 55
Trust, xiv, 63, 69, 79, 80, 97, 139, 148, 155, 189, 220, 221
oxytocin and, xiii, 59, 60–61
Trust game, 60, 69
Trust signals, 61, 79, 87, 108
Trut, Lyudmila, 76, 122–123
Tuan, Yi-Fu, 169
Tudge, Colin, 139, 140, 142
Twenge, Jean, 216

Umm Qasr, dolphins in, 237
U.S. Navy, marine mammal program of, 237
Uvnas-Moberg, Kerstin, 28, 72, 73
on engagement, 45
nurturing and, 188
oxytocin and, xi, xii, 55, 63, 186, 187, 206
petting and, 95–96
round-pens and, 92
SSRIs and, 191
stroking and, 95
vasopressin and, 63

Vasopressin, 48, 62, 93, 211
decrease in, 128
dominance of, 208
mating and, 22
oxytocin and, 21–22, 63, 65, 129, 175, 207
receptors, 22, 23, 24, 25, 58
Vervet monkeys, 113, 231–232
Violence, 85, 199
Viral infection, interspecies, 242
Virtual contact, 198–199
Visual signals, 2, 39, 93, 96, 106
Visualization, 101, 102

Vocalizations, 106, 108, 113, 114, 129
Voice
  intention of, 124
  tone/volume of, 107, 201, 203
Von Osten, Wilhelm, 88, 89, 91, 110
V1a vasopressin, 24–25, 63
V1aR allele, 62–63
V1aR receptors, 22, 23

War games, 199
Waters, Merian, 183–184
Weisman, Myrna, 196
Weitzenhoffer, Andre, 102
Welsch, Roger, 226
Wilderness, 239, 244
Wilkins, Gregory, 181, 183, 190, 192
Wilson, Edward O., 30, 31, 82
  biophilia and, 12–13, 14, 44
Wilson, Gilbert L., 140–141
Winnicott, Donald W., 29, 55
Witt, Dianne, 22
Wolf pups, breast-feeding, 72, 74, 131
Wolves, vii, 65–66, 216
  barking by, 127
  breeding, 122
  chemical shift in, xii–xiii
  cognitive capacities of, 123
  conversations with, 130
  cooperation among, 66, 69
  dogs and, xiv, 71, 124, 130, 131, 238
  humans and, 70, 75, 78, 106, 135–136, 159
  hunting tactics by, 69
  taming, 68, 235
  trust game and, 69
World Bank, 196
World Health Organization, 196
Wrangham, Richard, 122

York, Duke of, 192, 193
Young, Lawrence, 23, 24, 25

Zajonc, Robert, 6
Zak, Paul, 59–60
Zeki, Semir, 30
Zeuner, Frederick E., 68, 77, 148
Zoo program, 185, 187, 188, 189, 190, 191, 192
  ADHD and, 182–183
  assessment of, 183–184
  therapeutic effects of, 225
Zoos, as human-animal institutions, 239, 240–241

About the Author

Meg Daley has created and written cultural and historical documentaries for Emmy Award series such as *Smithsonian World, National Geographic Explorer, The Discovery Channel Specials,* and *PBS' The Living Edens.*

In 1992 while developing a series on the nature of the human-animal bond, she was asked to join a research team studying the neurobiology of social bonding headed by Dr. Carol Sue Carter of the University of Maryland and Dr. Kerstin Uvnas-Moberg of the Karolinska Institute in Stockholm. Despite her lack of formal scientific training, she made original and significant contributions to the understanding of the role social bonding plays in our mental and physical wellbeing. Her partnership in this scientific endeavor inspired *Made for Each Other.*

Meg Daley Olmert and her husband have a home on the eastern shore of Maryland which they share with their kayaking cats.